SCHAUM'S OUTLINE OF

THEORY AND PROBLEMS

of

BASIC MATHEMATICS
for Electricity and Electronics

ARTHUR BEISER, Ph.D.

SCHAUM'S OUTLINE SERIES
McGRAW-HILL, INC.

New York   St. Louis   San Francisco   Auckland   Bogotá   Caracas
Hamburg   Lisbon   London   Madrid   Mexico   Milan   Montreal
New Delhi   Paris   San Juan   São Paulo   Singapore
Sydney   Tokyo   Toronto

ARTHUR BEISER received his Ph.D. from New York University, where he subsequently served as Assistant and Associate Professor of Physics. He has been a consultant to various industrial firms and government agencies and is the author of more than a dozen textbooks of physics and mathematics.

Schaum's Outline of Theory and Problems of
BASIC MATHEMATICS FOR ELECTRICITY AND ELECTRONICS

9 10 11 12 13 14 15 SH SH 9 8 7 6 5 4 3 2 1 0

Sponsoring Editor, John Aliano
Editing Supervisor, Denise Schanck
Production Supervisor, Claudia Dukeshire

Library of Congress Cataloging in Publication Data

Beiser, Arthur.
   Schaum's outline of basic mathematics for electricity and electronics.

   (Schaum's outline series)
   Includes index.
   1. Electric engineering—Mathematics.   2.   Electronics—Mathematics.   I. Title.   II. Title: Mathematics for electricity and electronics.
TK153.B39   621.3'01'51   80-14316
ISBN 0-07-004378-7 (pbk.)

Cover design by Amy E. Becker.

Preface

The successful study of electrical and electronic fundamentals requires a certain command of mathematics. The best way to develop such a command is in the context of the subject matter itself, so that the mathematical methods are illustrated by calculations based upon actual problems. This approach, which is used here, not only helps the reader to learn the methods but also adds to his or her familiarity with the electrical principles involved.

The sequence of the electrical material has been chosen so that the mathematics involved starts with elementary algebra, goes on to such more advanced topics as logarithms and simultaneous equations, and finally considers the trigonometry required for working with alternating currents. All basic electrical calculations are covered, so the reader will be left in no doubt as to how to tackle any problem in this area. Because each topic is discussed as though brand-new to the reader, the book can be used to fill in gaps in his or her background as well as to expand that background to the extent required. Both British and SI (metric) units are covered, and powers-of-ten notation is emphasized. Although the book can serve as a supplement to any introductory text on electricity and electronics, it is sufficiently self-contained to be used by itself.

In addition to the sample problems that are solved in detail, supplementary problems are provided with only the answers given. Working out the supplementary problems will give the reader practice in using the mathematical methods they illustrate and will also reveal any imperfections in his or her understanding of them.

ARTHUR BEISER

Contents

CONTENTS

Chapter 1

Basic Electricity and Algebra

STRUCTURE OF MATTER

Elements are the fundamental substances of which matter in bulk is composed. There are 105 known elements, including a number not found in nature but which have been prepared in the laboratory. Two or more elements can combine to form a *compound*, which is a substance whose properties are different from those of its constituent elements. Thus the elements hydrogen and oxygen, which are gasses under ordinary conditions, are the constituents of the compound water, which is a liquid under these conditions.

The ultimate particles of an element are called *atoms*. An atom consists of a small, relatively heavy *nucleus* with a number of lighter *electrons* circling it some distance away. The nucleus in turn is composed of *protons* and *neutrons*, which, like electrons, are elementary particles that cannot be further subdivided.

Electric charge is a basic property of protons and electrons. There are two kinds of charge, positive charge, which is carried by protons, and *negative charge*, which is carried by electrons. (Neutrons have no charge.) Charges of the same sign repel each other, charges of opposite sign attract each other. The number of protons in the nucleus of an atom normally equals the number of electrons around it, so the atom as a whole is electrically neutral. The protons and neutrons in the nucleus are held together by a nonelectric force sufficiently strong to overcome the mutual repulsion of the protons. The forces between atoms that hold them together to form molecules, solids, and liquids are electric in origin.

ELECTRIC CURRENT

The unit of electric charge is the *coulomb* (C). All charges in nature occur in multiples of $\pm e = \pm 0.000\ 000\ 000\ 000\ 000\ 000\ 16$ C which is more conveniently written 1.6×10^{-19} C; see Chapter 4. The charge on the proton is $+e$ and that on the electron is $-e$.

A flow of charge from one place to another constitutes an *electric current*. The direction of a current is conventionally considered to be that in which a positive charge would have to move to produce the same effects as the actual current. Thus a current is always supposed to go from the positive terminal of a battery or generator to its negative terminal.

A *conductor* is a substance through which charge can flow easily and an *insulator* is one through which charge can flow only with great difficulty. Metals, many liquids, and plasmas (gases whose molecules are charged) are conductors; nonmetallic solids, certain liquids, and gases whose molecules are electrically neutral are insulators. A number of substances, called *semiconductors*, are intermediate in their ability to conduct charge.

Electric currents in metals always consist of flows of electrons; such currents are assumed to occur in the direction opposite to that in which the electrons move. Since a positive charge going one way is for most purposes equivalent to a negative charge going the other way, this assumption makes no practical difference. Both positive and negative charges move when a current is present in a liquid or gaseous conductor.

When an amount of charge Q passes a given point in a conductor in the time interval t, the current I in the conductor is

$$I = \frac{Q}{t}$$

$$\text{Current} = \frac{\text{charge}}{\text{time interval}}$$

1

The unit of electric current is the *ampere* (A), where

$$1 \text{ ampere} = 1 \frac{\text{coulomb}}{\text{second}}$$

OHM'S LAW

In order for water to flow through a pipe, a pressure must be applied to push it through. Similarly, an electrical equivalent of pressure is needed for charge to flow through a conductor. This equivalent is called *potential difference* (V). The unit of potential difference is the *volt* (V), and potential difference is often referred to simply as "voltage."

In the case of a metallic conductor, the current is proportional to the applied potential difference: doubling V causes I to double, tripling V causes I to triple, and so forth. This relationship is known as *Ohm's law* and is expressed in the form

$$I = \frac{V}{R}$$

$$\text{Current} = \frac{\text{potential difference}}{\text{resistance}}$$

The quantity R is a constant for a given conductor and is called its *resistance*. The unit of resistance is the *ohm* (Ω). The greater the resistance of a conductor, the less the current when a certain voltage is applied.

The resistance of a wire or other metallic conductor depends upon the material it is made of (a copper wire has only 1/7 the resistance of an iron wire whose size is the same); its length (the longer a wire is, the greater its resistance); and its cross-sectional area (the greater the cross-sectional area, the less the resistance).

Ohm's law is not a physical principle but is an experimental relationship that most metals obey over a wide range of values of V and I.

What the formula $I = V/R$ does is give a prescription for calculating the current I in terms of the voltage V and the resistance R in a circuit. The formula is not restricted to a particular circuit but can be applied to any circuit for which V and R are known. What is being expressed is the manner in which I varies with V and R: an increase in V means an increase in I in the same proportion (doubling V doubles I, for instance), and an increase in R means a decrease in I in the same proportion (doubling R halves I, for instance).

Problem 1.1.   A 120-V toaster has a resistance of 12 Ω.   Find the current in the toaster.

The voltage here is $V = 120$ V   and the resistance is $R = 12$ Ω.   Hence

$$I = \frac{V}{R} = \frac{120 \text{ V}}{12 \text{ }\Omega} = 10 \text{ A}$$

Note that volts divided by ohms equal amperes   ($\text{V}/\Omega = \text{A}$).   It is always wise to carry along units in a calculation as a check against mistakes.

Problem 1.2.   What would the current in the toaster in Problem 1.1 be if its resistance were 10 Ω? 15 Ω?

When $R = 10$ Ω,

$$I = \frac{V}{R} = \frac{120 \text{ V}}{10 \text{ }\Omega} = 12 \text{ A}$$

When $R = 15$ Ω,

$$I = \frac{V}{R} = \frac{120 \text{ V}}{15 \text{ }\Omega} = 8 \text{ A}$$

Reducing the resistance increases the current; increasing the resistance decreases the current.

Problem 1.3. What would the current in the toaster be if the applied voltage were 100 V?   150 V?

When $V = 100$ V,

$$I = \frac{V}{R} = \frac{100 \text{ V}}{12 \text{ } \Omega} = 8.3 \text{ A}$$

When $V = 150$ V,

$$I = \frac{V}{R} = \frac{150 \text{ V}}{12 \text{ } \Omega} = 12.5 \text{ A}$$

Reducing the voltage reduces the current; increasing the voltage increases the current.

What do we do if we know the current and resistance in a circuit and want to find the voltage? Or if we know the current and voltage and want to find the resistance? In order to answer such questions, and to be able to apply Ohm's law to more complicated circuits than those with only a single resistance and source of potential difference, a brief review of the elements of algebra is in order.

ALGEBRA

Algebra is the arithmetic of symbols that represent numbers. Instead of being restricted to relationships among specific numbers, algebra can express more general relationships among quantities whose numerical values need not be known.

The arithmetical operations of addition, subtraction, multiplication, and division have the same meaning in algebra. Addition and subtraction are straightforward:

$$x + y = a$$

means that we obtain the sum a by adding the quantity y to the quantity x, while

$$x - y = b$$

means that we obtain the difference b by subtracting the quantity y from the quantity x.

In algebraic multiplication, no special sign is ordinarily used, and the symbols of the quantities to be multiplied are merely written next to one another. Thus

$$xy = c$$

means the same as

$$x \times y = c$$

and

$$xyz = d$$

means the same as

$$x \times y \times z = d$$

When the quantity x is to be divided by y to yield the quotient e, we would write

$$\frac{x}{y} = e$$

which can also be expressed in the form

$$x/y = e$$

If several operations are to be performed in a certain order, parentheses ( ), brackets [ ], and braces { } are used to indicate this order. For instance, $a(x + y)$ means that we are first to add x and y together and then to multiply their sum $(x + y)$ by a. In essence $a(x + y)$ is an abbreviation for the same quantity written out in full:

$$a(x + y) = ax + ay$$

Problem 1.4. Remove the parentheses from $2a - 3(a + b) + 4(2a - b)$.

Since $3(a + b) = 3a + 3b$ and $4(2a - b) = 8a - 4b$,

$$2a - 3(a + b) + 4(2a - b) = 2a - 3a - 3b + 8a - 4b = (2 - 3 + 8)a - (3 + 4)b = 7a - 7b = 7(a - b)$$

Problem 1.5. Find the value of

$$v = 5\left(\frac{x - y}{z}\right) + w$$

when $x = 15$, $y = 3$, $z = 4$,   and $w = 10$.

We proceed as follows:

Step 1.   Subtract y from x to give

$$x - y = 15 - 3 = 12$$

Step 2.   Divide $x - y$ by z to give

$$\frac{x - y}{z} = \frac{12}{4} = 3$$

Step 3.   Multiply $(x - y)/z$ by 5 to give

$$5\left(\frac{x - y}{z}\right) = 5 \times 3 = 15$$

Step 4.   Add w to $5[(x - y)/z]$ to give

$$v = 5\left(\frac{x - y}{z}\right) + w = 15 + 10 = 25$$

POSITIVE AND NEGATIVE QUANTITIES

The rules for multiplying and dividing positive and negative quantities are simple. If the quantities are both positive or both negative, the result is positive; if one is positive and the other negative, the result is negative.   In symbolic form,

$$(+a) \times (+b) = (-a) \times (-b) = +ab \qquad (-a) \times (+b) = (+a) \times (-b) = -ab$$

$$\frac{+a}{+b} = \frac{-a}{-b} = +\frac{a}{b} \qquad\qquad \frac{-a}{+b} = \frac{+a}{-b} = -\frac{a}{b}$$

Problem 1.6. Examples of multiplication and division.

$$(-3) \times (-5) = 15 \qquad \frac{-16}{-4} = 4$$

$$2 \times (-4) = -8 \qquad \frac{10}{-5} = -2$$

$$(-12) \times 6 = -72 \qquad \frac{-24}{4} = -6$$

Problem 1.7. Find the value of

$$w = \frac{xy}{x + y}$$

when $x = 5$ and $y = -6$.

Here $xy = 5 \times (-6) = -30$ and $x + y = 5 + (-6) = 5 - 6 = -1$.   Hence

$$w = \frac{xy}{x + y} = \frac{-30}{-1} = 30$$

EQUATIONS

An equation is a statement of equality: whatever is on the left-hand side of an equal sign is equal to whatever is on the right-hand side.   An example of an arithmetical equation is

$$3 \times 9 + 8 = (3 \times 9) + 8 = 27 + 8 = 35$$

since it contains only numbers, and an example of an algebraic equation is

$$5x - 10 = 20$$

since it contains a symbol as well as numbers.

The symbols in an algebraic equation usually cannot have any arbitrary values if the equality is to hold.   To *solve* an equation is to find the possible values of these symbols.   The solution of the equation $5x - 10 = 20$ is $x = 6$ since only when $x = 6$ is this equation a true statement:

$$5x - 10 = 20$$
$$5 \times 6 - 10 = 20$$
$$30 - 10 = 20$$
$$20 = 20$$

The algebraic procedures that can be used to solve an equation are all based on the principle that any operation performed on one side of an equation must be performed on the other side as well. Thus an equation remains valid when the same quantity is added to or subtracted from both sides or is used to multiply or divide both sides.

Two helpful rules follow from the above principle.   The first is that any term on one side of an equation may be transposed to the other side by changing its sign.   To verify this rule, we consider the equation

$$a + b = c$$

If we subtract b from each side of the equation, we obtain

$$a + b - b = c - b$$
$$a = c - b$$

Thus b has disappeared from the left-hand side and $-b$ is now on the right-hand side.   Similarly, if

$$a - d = e$$

then it is true that

$$a = e + d$$

The second rule is that a quantity that multiplies one side of an equation may be transposed so as to divide the other side, and vice versa.   To verify this rule, we consider the equation

$$ab = c$$

If we divide both sides of the equation by b, we obtain

$$\frac{ab}{b} = \frac{c}{b}$$

The b's on the left cancel out to leave

$$a = \frac{c}{b}$$

Thus b, which was a multiplier on the left-hand side, is now a divisor on the right-hand side. Similarly, if

$$\frac{a}{d} = e$$

then it is true that

$$a = ed$$

Problem 1.8.   Solve $5x - 10 = 20$ for the value of x.

What we want to do is have just x on the left-hand side of the equation. The first step is to transpose the -10 to the right-hand side, where it becomes $+10$:

$$5x - 10 = 20$$
$$5x = 20 + 10 = 30$$

Now we transpose the 5 so that it divides the right-hand side:

$$5x = 30$$
$$x = \frac{30}{5} = 6$$

The solution is $x = 6$.

Problem 1.9. Solve Ohm's law for V and for R.

Ohm's law states that

$$I = \frac{V}{R}$$

To solve for V, we bring the R to the left-hand side, where it multiplies I:

$$IR = V$$

We now reverse the equation to read

$$V = IR$$

To solve for R, we start from $IR = V$ and transpose the I to the right-hand side, where it divides V:

$$IR = V$$
$$R = \frac{V}{I}$$

Thus Ohm's law can be expressed in three ways, depending upon which quantities are known and which are to be found:

$$I = \frac{V}{R} \qquad V = IR \qquad R = \frac{V}{I}$$

In using these formulas, it is necessary to keep in mind that

$$\text{Amperes} = \frac{\text{volts}}{\text{ohms}} \qquad \text{Volts} = \text{amperes} \times \text{ohms} \qquad \text{Ohms} = \frac{\text{volts}}{\text{amperes}}$$

$$A = \frac{V}{\Omega} \qquad V = A \times \Omega \qquad \Omega = \frac{V}{A}$$

Problem 1.10. The current in the coil of an 8-Ω loudspeaker is 2 A. Find the potential difference across its terminals.

$$V = IR = 2\ A \times 8\ \Omega = 16\ V$$

Problem 1.11. A 120-V electric heater draws a current of 25 A. What is its resistance?

$$R = \frac{V}{I} = \frac{120\ V}{25\ A} = 4.8\ \Omega$$

CROSS MULTIPLICATION

When each side of an equation consists of a fraction, all we need do to remove the fractions is *cross multiply*:

$$\frac{a}{b} = \frac{c}{d} \longrightarrow ad = bc$$

What was originally the denominator (lower part) of each fraction now multiplies the numerator (upper part) of the other side of the equation.

Problem 1.12. Examples of cross multiplication.

$$\frac{x}{2} = \frac{y}{7} \qquad\qquad \frac{y}{8} = \frac{5}{x}$$
$$7x = 2y \qquad\qquad xy = 5 \times 8 = 40$$

$$\frac{3x}{5} = \frac{4y}{3} \qquad\qquad \frac{y}{5} = \frac{3x+2}{4}$$
$$3(3x) = 5(4y) \qquad 4y = 5(3x+2) = 15x + 10$$
$$9x = 20y$$

Problem 1.13. Solve the equation

$$\frac{5}{a+2} = \frac{3}{a-2}$$

for the value of a.

We proceed as follows:

Cross multiply to give	$5(a-2) = 3(a+2)$
Multiply out both sides to give	$5a - 10 = 3a + 6$
Transpose the -10 and the $3a$ to give	$5a - 3a = 6 + 10$
Carry out the indicated addition and subtraction to give	$2a = 16$
Divide both sides by 2 to give	$a = 8$

Problem 1.14. Solve the equation

$$\frac{16x - 2}{8} = \frac{3x}{1}$$

for the value of x.

Cross multiply to give	$16x - 2 = 8(3x) = 24x$
Transpose the -2 and $24x$ to give	$16x - 24x = 2$
Combine like terms to give	$-8x = 2$
Divide both sides by -8 to give	$x = -1/4 = 0.25$

Problem 1.15. Solve the equation

$$\frac{4x - 35}{3} = 9(1 - x)$$

for the value of x.

Cross multiply to give	$4x - 35 = 3 \times 9(1-x) = 27(1-x)$
Multiply out the right-hand side to give	$4x - 35 = 27 - 27x$
Transpose the -35 and $-27x$ to give	$4x + 27x = 27 + 35$
Combine like terms to give	$31x = 62$
Divide both sides by 31 to give	$x = 62/31 = 2$

WHEATSTONE BRIDGE

The *Wheatstone bridge* (Fig. 1-1) provides an accurate means for determining an unknown resistance R_x with the help of the fixed resistors R_1 and R_2 and the calibrated variable resistor R_3. The resistance R_3 is varied until the galvanometer G shows no deflection, a situation described by saying that the bridge is *balanced*. (A galvanometer is a very sensitive instrument for measuring current.)

No current passes through the galvanometer when the bridge is balanced, which means that the same current I_1 passes through R_1 and R_2 and the same current I_2 passes through R_3 and R_x. Also, points A and B must have no potential difference between them, hence $V_{AC} = V_{BC}$ and $V_{AD} = V_{BD}$. Since

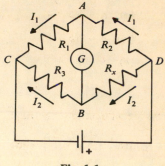

$$V_{AC} = I_1 R_1 \qquad V_{AD} = I_1 R_2$$
$$V_{BC} = I_2 R_3 \qquad V_{BD} = I_2 R_x$$

we have the equations

$$V_{AC} = V_{BC} \qquad V_{AD} = V_{BD}$$
$$I_1 R_1 = I_2 R_3 \qquad I_1 R_2 = I_2 R_x$$

Fig. 1-1

Dividing the last equation in the first column by the last equation in the second column gives

$$\frac{I_1 R_1}{I_1 R_2} = \frac{I_2 R_3}{I_2 R_x}$$

The currents I_1 and I_2 cancel out, and so

$$\frac{R_1}{R_2} = \frac{R_3}{R_x}$$

Since R_1, R_2, and R_3 are known, the value of R_x can be calculated. In most Wheatstone bridges the ratio R_2/R_1 can be adjusted to give ratios of $1:1$, $1:10$, $1:100$, and so forth to reduce the range required for the variable resistor R_3.

Problem 1.16. A Wheatstone bridge is balanced when $R_1 = 10 \ \Omega$, $R_2 = 1000 \ \Omega$, and $R_3 = 26 \ \Omega$. Find the unknown resistance.

From the equation of balance we cross-multiply to obtain

$$R_x = \frac{R_2 R_3}{R_1}$$

Substituting the known resistances yields

$$R_x = \frac{1000 \ \Omega \times 26 \ \Omega}{10 \ \Omega} = 2600 \ \Omega$$

Supplementary Problems

1.17. Remove the parentheses from the following and combine like terms:

(a) $a + (b + c - d)$ (f) $b - 3(b + 3)$

(b) $a - (b + c - d)$ (g) $2a - 3b - 4(a - 2b)$

(c) $a + 2(b - 4)$ (h) $3(a + b) - 3(a + 2b)$

(d) $a - 2(b - 4)$ (i) $2(a + b) - 3(a - b) + 4(a + 2b - c)$

(e) $b + 3(b + 3)$ (j) $5(a + b + c) - 5(a - b - c)$

1.18. Evaluate the following:

(a) $\dfrac{3(x + y)}{2}$ when $x = 5$ and $y = -2$

(b) $\dfrac{1}{x - y} - \dfrac{1}{x + y}$ when $x = 3$ and $y = 2$

(c) $\dfrac{4xy}{y + 3x} + 5$ when $x = 1$ and $y = -2$

(d) $\dfrac{x + y}{2z} + \dfrac{z}{x - y}$ when $x = -2,\ y = 2,$ and $z = 4$

(e) $\dfrac{x + z}{y} - \dfrac{xy}{2}$ when $x = 2,\ y = 8,$ and $z = 10$

(f) $\dfrac{3(x + 7)}{y + 2}$ when $x = 3$ and $y = -6$

(g) $\dfrac{5(3 - x)}{2(x + y)}$ when $x = -5$ and $y = 7$

1.19. Solve each of the following equations for x:

(a) $3x + 7 = 13$ (e) $\dfrac{x + 7}{6} = x + 2$ (i) $\dfrac{3}{x - 1} = \dfrac{5}{x + 1}$

(b) $5x - 8 = 17$ (f) $\dfrac{4x - 35}{3} = 9(1 - x)$ (j) $\dfrac{1}{3x + 4} = \dfrac{2}{x + 8}$

(c) $2(x + 5) = 6$ (g) $\dfrac{3x - 42}{9} = 2(7 - x)$ (k) $\dfrac{8}{x} = \dfrac{1}{4 - x}$

(d) $7x - 10 = 0.5$ (h) $\dfrac{1}{x + 1} = \dfrac{1}{2x - 1}$ (l) $\dfrac{x}{2x - 1} = \dfrac{5}{7}$

1.20. Find the resistance of a 240-V electric stove that draws a current of 12 A.

1.21. Find the resistance of a 120-V electric iron that draws a current of 15 A.

1.22. A 240-V water heater has a resistance of 24 Ω. Find the minimum rating of the fuse in the heater circuit.

1.23. What potential difference must be applied across a 1200-Ω resistor in order to produce a current of 0.05 A?

1.24. What potential difference must be applied across the above resistor in order to produce a current twice as great, namely 0.1 A?

1.25. A potential difference of 6 V is applied across a 40-Ω resistor.   What is the resulting current?

1.26. A Wheatstone bridge is balanced when $R_1 = 100\ \Omega$, $R_2 = 10\ \Omega$, and $R_3 = 8.26\ \Omega$. Find the unknown resistance.

Answers to Supplementary Problems

1.17. (a) $a + b + c - d$ (e) $4b + 9$ (h) $-3b$
    (b) $a - b - c + d$ (f) $-2b - 9$ (i) $3a + 13b - 4c$
    (c) $a + 2b - 8$ (g) $-2a + 5b$ (j) $10b + 10c$
    (d) $a - 2b + 8$

1.18. (a) 4.5 (b) 0.8 (c) -3 (d) -1 (e) -6.5 (f) -7.5 (g) 10

1.19. (a) 2   (d) 1.5   (g) 8   (j) 0
    (b) 5   (e) -1 (h) 2   (k) 32/9
    (c) -2 (f) 2   (i) 4   (l) 5/3

1.20. 20 Ω   **1.24.** 120 V

1.21. 8 Ω   **1.25.** 0.15 A

1.22. 10 A   **1.26.** 0.826 Ω

1.23. 60 V

Chapter 2

Fractions, Decimals, and Percentage

FRACTIONS

A *fraction* is some part of a total quantity. If we cut a pie into five equal portions, each portion is 1/5 of the entire pie. Every fraction consists of a number called the *numerator* which is divided by another number called the *denominator*:

$$\text{Fraction} = \frac{\text{numerator}}{\text{denominator}}$$

In the case of 1/5, the numerator is 1 and the denominator is 5.

Any fraction can be expressed in different ways. Thus 2/10 is really the same as 1/5 because both represent the same proportion of the total quantity. They are called equivalent fractions. Other fractions equivalent to 1/5 are 3/15, 4/20, 10/50, 25/125, and so on. In every case the denominator is exactly five times the numerator. To *reduce a fraction to lowest terms* means to simplify it so that the numerator and denominator have no common factors. The common factor in 2/10 is 2, and to reduce 2/10 to lowest terms involves dividing numerator and denominator by 2:

$$\frac{2}{10} = \frac{2/2}{10/2} = \frac{1}{5}$$

Similarly the common factor in 3/15 is 3, so that

$$\frac{3}{15} = \frac{3/3}{15/3} = \frac{1}{5}$$

is this fraction in its lowest terms.

Problem 2.1. Express 15/40 and 12/64 in lowest terms.

Since 40 cannot be evenly divided by 15, this is not a common factor. However, both 15 and 40 are divisible by 5, and so

$$\frac{15}{40} = \frac{15/5}{40/5} = \frac{3}{8}$$

In the case of 12/64, both 12 and 64 are evenly divisible by 4 and by 2. Since 4 is the larger factor, we must use it to reduce 12/64 to lowest terms:

$$\frac{12}{64} = \frac{12/4}{64/4} = \frac{3}{16}$$

IMPROPER FRACTIONS

A *proper fraction* represents a quantity between 0 and 1, so that the numerator is smaller than the denominator; 1/5 and 3/16 are proper fractions. An *improper fraction* has a value greater than 1, so that the numerator is larger than the denominator; 6/5 and 42/12 are improper fractions. A *mixed number* consists of a whole number and a proper fraction, for instance $2\frac{3}{4}$ and $7\frac{1}{8}$. Every improper fraction can be expressed as a mixed number and vice-versa.

Problem 2.2. Change 6/5 and 42/12 into mixed numbers.

The procedure is to divide the numerator by the denominator and show the remainder divided by the denominator. Thus

$$\frac{6}{5} = 1\tfrac{1}{5}$$

because 5 fits into 6 with 1 left over:

$$\frac{6}{5} = \frac{5}{5} + \frac{1}{5} = 1\tfrac{1}{5}$$

In the same way

$$\frac{42}{12} = 3\tfrac{6}{12} = 3\tfrac{1}{2}$$

because

$$\frac{42}{12} = \frac{36}{12} + \frac{6}{12} = 3\tfrac{6}{12}$$

Since $6/12 = 1/2$, the answer is properly expressed as $3\tfrac{1}{2}$.

Problem 2.3. Change $2\tfrac{3}{4}$ and $7\tfrac{1}{8}$ into improper fractions.

Here we multiply the whole number by the denominator of the fraction and add the result to its numerator to give the numerator of the improper fraction. The denominator remains the same. Thus

$$2\tfrac{3}{4} = \frac{(2 \times 4) + 3}{4} = \frac{8+3}{4} = \frac{11}{4} \qquad 7\tfrac{1}{8} = \frac{(7 \times 8) + 1}{8} = \frac{56+1}{8} = \frac{57}{8}$$

ADDITION AND SUBTRACTION

To add or subtract two or more fractions, they must have the same denominator. The sum of the fractions is just the sum of their numerators over the same denominator:

$$\frac{1}{3} + \frac{1}{3} = \frac{1+1}{3} = \frac{2}{3} \qquad \frac{1}{8} + \frac{3}{8} = \frac{1+3}{8} = \frac{4}{8} = \frac{1}{2} \qquad \frac{1}{8} + \frac{3}{8} + \frac{5}{8} = \frac{1+3+5}{8} = \frac{9}{8} = 1\tfrac{1}{8}$$

If the denominators are not the same, a common one must be found and the fractions changed to this denominator first:

$$\frac{1}{4} + \frac{1}{8} = \frac{2}{8} + \frac{1}{8} = \frac{2+1}{8} = \frac{3}{8} \qquad \frac{1}{2} + \frac{7}{16} = \frac{8}{16} + \frac{7}{16} = \frac{8+7}{16} = \frac{15}{16}$$

Problem 2.4. Add 2/5 and 1/3.

Here both fractions must be changed. Since one has the denominator 5 and the other has the denominator 3, their product $5 \times 3 = 15$ is a suitable common denominator. We have

$$\frac{2}{5} = \frac{2 \times 3}{5 \times 3} = \frac{6}{15} \qquad \text{and} \qquad \frac{1}{3} = \frac{1 \times 5}{3 \times 5} = \frac{5}{15}$$

and so

$$\frac{2}{5} + \frac{1}{3} = \frac{6}{15} + \frac{5}{15} = \frac{11}{15}$$

Subtraction follows the same pattern as addition:

$$\frac{7}{8} - \frac{1}{8} = \frac{7-1}{8} = \frac{6}{8} = \frac{3}{4} \qquad \frac{3}{4} - \frac{1}{2} = \frac{3}{4} - \frac{2}{4} = \frac{3-2}{4} = \frac{1}{4} \qquad \frac{3}{16} - \frac{9}{32} = \frac{6}{32} - \frac{9}{32} = -\frac{3}{32}$$

Mixed numbers can be converted into improper fractions with the same denominator in order to be added or subtracted:

$$2\tfrac{3}{4} + \frac{3}{4} = \frac{11}{4} + \frac{3}{4} = \frac{14}{4} = 3\tfrac{2}{4} = 3\tfrac{1}{2} \qquad 4\tfrac{1}{8} - 1\tfrac{7}{8} = \frac{33}{8} - \frac{15}{8} = \frac{18}{8} = 2\tfrac{2}{8} = 2\tfrac{1}{4}$$

Problem 2.5.   Subtract $1/4$ from $2\frac{2}{3}$.

A suitable common denominator is $4 \times 3 = 12$.   Hence

$$2\frac{2}{3} - \frac{1}{4} = \frac{8}{3} - \frac{1}{4} = \frac{8 \times 4}{3 \times 4} - \frac{1 \times 3}{4 \times 3} = \frac{32}{12} - \frac{3}{12} = \frac{29}{12} = 2\frac{5}{12}$$

MULTIPLICATION

To multiply a fraction by a whole number, multiply the numerator of the fraction by the whole number and put the result over the denominator.   For example,

$$3 \times \frac{1}{4} = \frac{3 \times 1}{4} = \frac{3}{4} \qquad 10 \times \frac{2}{3} = \frac{10 \times 2}{3} = \frac{20}{3} = 6\frac{2}{3}$$

To multiply a mixed number by a whole number, the mixed number must first be changed to an improper fraction.

Problem 2.6.   An electric drill requires a current of $2\frac{2}{3}$ A.   How much current is required by four such drills?

$$4 \times 2\frac{2}{3} \text{A} = 4 \times \frac{8}{3} \text{A} = \frac{32}{3} \text{A} = 10\frac{2}{3} \text{A}$$

Two or more fractions are multiplied together by multiplying their numerators to get the numerator of the product and multiplying their denominators to get the denominator of the product.   Thus

$$\frac{5}{7} \times \frac{1}{3} = \frac{5 \times 1}{7 \times 3} = \frac{5}{21}$$

$$\frac{3}{4} \times \frac{1}{3} \times \frac{2}{5} = \frac{3 \times 1 \times 2}{4 \times 3 \times 5} = \frac{6}{60} = \frac{1}{10}$$

$$1\frac{1}{2} \times 2\frac{2}{3} = \frac{3}{2} \times \frac{8}{3} = \frac{3 \times 8}{2 \times 3} = \frac{24}{6} = 4$$

When the fractions to be multiplied have the same number in the numerator of one and in the denominator of the other, this number can be canceled out to save some arithmetic.   Thus in the last example above, the 3s cancel each other because $3/3 = 1$:

$$1\frac{1}{2} \times 2\frac{2}{3} = \frac{\overset{1}{\cancel{3}}}{2} \times \frac{8}{\underset{1}{\cancel{3}}} = \frac{8}{2} = 4$$

A similar procedure is followed when the numerator of one fraction and the denominator of the other have a common factor.   Here the common factor is 5:

$$\frac{1}{5} \times 3\frac{1}{3} = \frac{1}{5} \times \frac{10}{3} = \frac{1}{\underset{1}{\cancel{5}}} \times \frac{\overset{2}{\cancel{10}}}{3} = \frac{2}{3}$$

Problem 2.7.   A variable resistor has a resistance of $7\frac{1}{2}$ Ω/in. What is the resistance when the sliding contact is $1\frac{1}{3}$ in from the end of the resistor?

$$1\frac{1}{3} \text{in} \times 7\frac{1}{2} \frac{\Omega}{\text{in}} = \frac{4}{3} \text{in} \times \frac{15}{2} \frac{\Omega}{\text{in}}$$

Dividing the 15 by 3 and the 4 by 2, we have

$$\frac{\overset{2}{\cancel{4}}}{\underset{1}{\cancel{3}}} \text{ in} \times \frac{\overset{5}{\cancel{15}}}{\underset{1}{\cancel{2}}} \frac{\Omega}{\text{in}} = 10 \ \Omega$$

Note that "in" in numerator and denominator cancel out just as the numbers do. The answer is therefore 10 Ω.

DIVISION

The *reciprocal* of a number is 1 divided by that number. Thus the reciprocal of 5 is 1/5. In the case of a fraction, the reciprocal is formed by inverting it, that is, by exchanging its numerator and denominator. Thus the reciprocal of 2/3 is 3/2.

The process of dividing one number by another is the same as multiplying the first number by the reciprocal of the second. For instance, to divide 15 by 5 is the same as multiplying 15 by 1/5:

$$\frac{15}{5} = 15 \times \frac{1}{5} = 3$$

The same procedure is followed in the case of fractions. Here are some examples:

$$\frac{5/8}{3} = \frac{5}{8} \times \frac{1}{3} = \frac{5 \times 1}{8 \times 3} = \frac{5}{24} \qquad \frac{4}{3/4} = 4 \times \frac{4}{3} = \frac{4 \times 4}{3} = \frac{16}{3} = 5\frac{1}{3}$$

Note here that 4/3 is the reciprocal of 3/4.

$$\frac{2/5}{3/8} = \frac{2}{5} \times \frac{8}{3} = \frac{2 \times 8}{5 \times 3} = \frac{16}{15} = 1\frac{1}{15}$$

Here 8/3 is the reciprocal of 3/8.

When a fraction is divided by a whole number, it is easiest to just multiply the denominator of the fraction by the number. To divide 5/8 by 3, we need only write

$$\frac{5/8}{3} = \frac{5}{8 \times 3} = \frac{5}{24}$$

In the case of division involving a mixed number, the number must first be changed to an improper fraction:

$$\frac{2\frac{1}{4}}{3} = \frac{9/4}{3} = \frac{9}{4 \times 3} = \frac{9}{12} = \frac{3}{4}$$

$$\frac{6}{1\frac{1}{2}} = \frac{6}{3/2} = 6 \times \frac{2}{3} = \frac{6 \times 2}{3} = \frac{12}{3} = 4$$

$$\frac{5\frac{5}{8}}{2\frac{1}{4}} = \frac{45/8}{9/4} = \frac{\overset{5}{\cancel{45}}}{\underset{2}{\cancel{8}}} \times \frac{\overset{1}{\cancel{4}}}{\underset{1}{\cancel{9}}} = \frac{5}{2} = 2\frac{1}{2}$$

DECIMALS

A decimal fraction always has as its denominator a number from the sequence 10, 100, 1000, 10 000, ... The value of the denominator is not written but is indicated by the position of the decimal point in the numerator of the fraction. If only one digit follows the decimal point, the denominator is 10; if two digits follow the decimal point, the denominator is 100; if three digits follow the decimal point, the denominator is 1000; and so on. Here are some examples:

$$\frac{3}{10} = 0.3 \qquad\qquad \frac{8}{1000} = 0.008$$

$$\frac{5}{100} = 0.05 \qquad\qquad \frac{95}{1000} = 0.095$$

$$\frac{64}{100} = 0.64 \qquad\qquad \frac{222}{1000} = 0.222$$

Improper fractions with 10, 100, 1000, . . . in their denominators follow the same pattern:

$$\frac{12}{10} = 1.2 \qquad\qquad \frac{431}{100} = 4.31$$

$$\frac{859}{10} = 85.9 \qquad\qquad \frac{9403}{1000} = 9.403$$

A zero at the right end of a decimal numeral does not affect its value but may be useful to indicate the precision with which the number is known or to help in comparing it with another number. For instance, if a measurement of the diameter of a shaft is reported as 0.25 in, all we are entitled to conclude is that the actual diameter is somewhere between 0.245 and 0.255 in, a range of 0.01 in. If instead the diameter is said to be 0.250 in, the uncertainty in the figure is reduced by a factor of 1/10 and the actual diameter must be between 0.2495 and 0.2505 in.

If we want to compare two decimals, for instance 0.7 and 0.63, it helps to think of 0.7 as 0.70. Since 70/100 is greater than 63/100, 0.7 is larger than 0.63.

To change a fraction into its decimal equivalent, all that is necessary is to divide the numerator by the denominator.

Problem 2.8. Find the decimal equivalent of 1/2.

To carry out the division of 1 by 2 it is necessary first to put a decimal point after the 1 and add a zero. This operation does not change the value of the 1 since   1.0 = 1.   We have

$$\begin{array}{r} 0.5 \\ 2\,\overline{\big)\,1.0} \end{array}$$

and so 1/2 = 0.5.

Problem 2.9. Find the decimal equivalent of 5/16.

Now we need a decimal point and four zeros after the 5 in order to have no remainder:

$$
\begin{array}{r}
0.3125 \\
16\,\overline{\big)\,5.0000} \\
4.8 \quad\;\; \\
\hline
20 \;\;\; \\
16 \;\;\; \\
\hline
40 \;\; \\
32 \;\; \\
\hline
80 \\
80 \\
\hline
0 \\
\end{array}
$$

Hence   5/16 = 0.3125.   We can add as many zeros as required.

In the case of a fraction like 1/3, where   1/3 = 0.33333333 . . . ,   we keep as many digits as the situation requires.   Depending on circumstances, we might use 0.33, 0.333, or 0.3333 for the value of 1/3.   The convention for rounding off is that if the digit after the last one to be kept is 4 or less, it is simply dropped, whereas if it is 5 or more, the last digit to be kept is increased by 1.   Thus 0.484 is rounded off to 0.48 but 0.485 is rounded off to 0.49.

ARITHMETIC WITH DECIMALS

To add decimals, it is important that their decimal points be in line. Then all that is necessary is to add the digits in each column in the usual way, considering missing digits to be zero.   Thus 47 + 1.57 + 0.031   would be worked out in the form

$$
\begin{array}{r}
47 \\
1.57 \\
+\ 0.031 \\
\hline
48.601
\end{array}
$$

In subtraction, it is helpful to add any zeros required to give the same number of decimal places in both numbers.   Thus   28.3 − 5.81   would be worked out in the form

$$
\begin{array}{r}
28.30 \\
-\ 5.81 \\
\hline
22.49
\end{array}
$$

When decimals are multiplied, the number of decimal places in the product is equal to the sum of the number of decimal places in the original numbers.

Problem 2.10.   A battery provides a current of 3.7 A for 1.4 hours (h).   How much energy in A · h did the battery supply?

The product of 37 and 14 is 518.   Since 3.7 and 1.4 each have one decimal place, their product must have two decimal places.   Hence

$$3.7 \text{ A} \times 1.4 \text{ h} = 5.18 \text{ A} \cdot \text{h}$$

Problem 2.11.   Number 16 copper wire has a resistance of 0.004 Ω/ft.   What is the resistance of 0.96 ft of this wire?

The product of 96 and 4 is 384.   Since 0.96 has two decimal places and 0.004 has three, their product must have five decimal places.   In order to achieve this result, two zeros must be placed between the decimal point and the 384:

$$0.96 \text{ ft} \times 0.004 \text{ }\Omega/\text{ft} = 0.00384 \text{ }\Omega$$

Note that   ft $\times$ Ω/ft = Ω since ft cancel out.

Problem 2.12.   A certain type of six-conductor cable weighs 4.25 lb per 100 ft.   How much does 272 ft of this cable weigh?

The weight of the cable per foot is

$$\frac{4.25 \text{ lb}}{100 \text{ ft}} = 0.0425 \ \frac{\text{lb}}{\text{ft}}$$

Hence 272 ft of the cable weighs

$$272 \text{ ft} \times 0.0425 \text{ lb/ft} = 11.5600 \text{ lb}$$

The final two zeros are present because 0.0425 has four decimal places.   Since these zeros serve no purpose here, they can be omitted and the answer given as 11.56 lb.

To divide a decimal by a whole number, the usual long-division procedure is followed. Thus 1.848 divided by 22 is carried out by hand in this way:

$$\begin{array}{r} 0.084 \\ 22\overline{)1.848} \\ 1.76 \\ \hline 88 \\ 88 \\ \hline 0 \end{array}$$

When a decimal is the divisor, the most convenient thing to do is to shift the decimal point in both numbers by the same number of places to the right so the divisor is a whole number.

Problem 2.13. Divide 5.04 by 0.08.

$$0.08\overline{)5.04} = 8\overline{)504} \quad \begin{array}{r} 63 \\ \\ 48 \\ \hline 24 \\ 24 \\ \hline 0 \end{array}$$

The answer is 63.

Problem 2.14. Divide 1.6 by 0.64.

Here an extra zero must be added to the 1.6 in order to shift its decimal point the required two places to the right. In the course of the calculation, another zero will have to be added to complete the process:

$$0.64\overline{)1.60} = 64\overline{)160.0} \quad \begin{array}{r} 2.5 \\ \\ 128 \\ \hline 320 \\ 320 \\ \hline 0 \end{array}$$

The answer is 2.5.

PERCENT

A quantity is often expressed as a *percentage* of another quantity of the same kind, where

$$1 \text{ percent} = 1\% = \frac{1}{100} = 0.01$$

Thus

$$100\% = \frac{100}{100} = 1 \qquad 35\% = \frac{35}{100} = 0.35 \qquad 0.2\% = \frac{0.2}{100} = 0.002 \qquad 160\% = \frac{160}{100} = 1.6$$

Evidently a percentage can be changed to its decimal equivalent by shifting the decimal point two places to the left and deleting the percent sign.

To change a decimal to a percentage, shift the decimal point two places to the right and add a percent sign.   Thus

$$0.78 = 78\% \qquad 0.2364 = 23.64\% \qquad 8.7 = 870\%$$

To change a fraction into a percentage, first express it as a decimal.   Then shift the decimal point two places to the right and add a percent sign.   Thus

$$\frac{1}{2} = 0.5 = 50\% \qquad \frac{3}{4} = 0.75 = 75\% \qquad 1\tfrac{5}{16} = \frac{21}{16} = 1.3125 = 131.25\%$$

Problem 2.15.   A sales tax of 6% is applied to the purchase of a $50 drill.   What is the amount of the tax?

Since   6% = 0.06,   tax = 0.06 × $50 = $3.

Problem 2.16.   A discount of 25% is allowed on the purchase of parts whose total list price is $150. What is the net price?

Since   25% = 0.25,

$$\text{Discount} = 0.25 \times \$150 = \$37.50$$

The net price is therefore

$$\text{Net price} = \text{list price} - \text{discount} = \$150 - \$37.50 = \$112.50$$

Problem 2.17.   A person whose salary is $300 per week is given a raise of $36 per week.   What percentage of his original salary is this?

$$\frac{\text{Salary increase}}{\text{Original salary}} = \frac{\$36}{\$300} = 0.12 = 12\%$$

Problem 2.18.   The voltage drop in a transmission line is 4.8 V.   If this is 2% of the source voltage, find (*a*) the source voltage and (*b*) the voltage at the end of the transmission line.

(*a*)   Since   voltage drop = fraction lost × source voltage   and   2% = 0.02,

$$\text{Source voltage} = \frac{\text{voltage drop}}{\text{fraction lost}} = \frac{4.8 \text{ V}}{0.02} = 240 \text{ V}$$

(*b*)

$$\text{Terminal voltage} = \text{source voltage} - \text{voltage drop} = 240 \text{ V} - 4.8 \text{ V} = 235.2 \text{ V}$$

Problem 2.19.   A certain fuse rated at 15 A requires an actual current of 16 A to blow.   What percentage of its rating is this?

$$\frac{\text{Actual current}}{\text{Rated current}} = \frac{16 \text{ A}}{15 \text{ A}} = 1.07 = 107\%$$

Problem 2.20.   A gold fourth band on a resistor indicates that its actual value is within ±5% of the indicated value.   What is the range of possible values of a 3900-Ω resistor which has such a gold band?   Remember that   ±5% = ±0.05.

The highest deviation of the resistor from the nominal value of 3900 Ω is

$$0.05 \times 3900 \text{ } \Omega = 195 \text{ } \Omega$$

Hence its actual resistance may be anywhere from a minimum of

$$R_{\min} = (3900 - 195) \text{ } \Omega = 3705 \text{ } \Omega$$

to a maximum of

$$R_{\max} = (3900 + 195) \text{ } \Omega = 4095 \text{ } \Omega$$

Problem 2.21. What is the efficiency of a generator that delivers 2.5 kW of electric power when it is supplied with 2.9 kW of mechanical power?

In general,

$$\text{Efficiency} = \frac{\text{output}}{\text{input}}$$

Here

$$\text{Efficiency} = \frac{2.5 \text{ kW}}{2.9 \text{ kW}} = 0.86 = 86\%$$

Problem 2.22. A motor is supplied with 500 W of electric power. If its efficiency is 80%, what is its output of mechanical power?

Here the efficiency is   80% = 0.80   and the input is 500 W.   Hence

$$\text{Output} = \text{efficiency} \times \text{input} = 0.80 \times 500 \text{ W} = 400 \text{ W}$$

Problem 2.23. A transformer delivers 1200 W of electric power. If its efficiency is 97%, find its power input.

$$\text{Input} = \frac{\text{output}}{\text{efficiency}} = \frac{1200 \text{ W}}{0.97} = 1237 \text{ W}$$

Supplementary Problems

2.24. Express the following fractions in lowest terms:

(a) 9/15 (b) 12/72 (c) 14/16 (d) 25/60 (e) 3/21 (f) 48/100

2.25. Express the following fractions as mixed numbers:

(a) 12/4 (b) 8/3 (c) 17/8 (d) 95/8 (e) 60/16 (f) 48/20

2.26. Express the following mixed numbers as fractions:

(a) $1\frac{1}{2}$ (b) $7\frac{1}{3}$ (c) $3\frac{5}{8}$ (d) $2\frac{3}{10}$ (e) $14\frac{5}{6}$ (f) $1\frac{15}{16}$

2.27. Perform the indicated additions:

(a) $\frac{2}{3} + \frac{1}{3}$ (c) $\frac{5}{6} + \frac{2}{3}$ (e) $\frac{7}{12} + \frac{1}{3} + \frac{4}{5}$ (g) $2\frac{1}{8} + \frac{3}{4}$

(b) $\frac{7}{8} + \frac{5}{8}$ (d) $\frac{1}{4} + \frac{2}{5}$ (f) $1\frac{1}{4} + \frac{3}{4}$ (h) $4\frac{2}{3} + 2\frac{1}{2}$

2.28. Perform the indicated subtractions:

(a) $\frac{2}{3} - \frac{1}{3}$ (c) $\frac{5}{6} - \frac{2}{3}$ (e) $4\frac{5}{8} - \frac{7}{8}$ (g) $10 - 2\frac{2}{5}$

(b) $\frac{7}{8} - \frac{5}{8}$ (d) $\frac{4}{5} - \frac{1}{4}$ (f) $2\frac{1}{3} - \frac{3}{4}$ (h) $1\frac{1}{4} - 2\frac{1}{8}$

2.29. Perform the indicated multiplications:

(a) $2 \times \frac{7}{8}$ (c) $\frac{1}{4} \times \frac{1}{4}$ (e) $1\frac{1}{2} \times \frac{7}{8}$ (g) $2\frac{5}{8} \times 1\frac{1}{3}$

(b) $5 \times 2\frac{4}{7}$ (d) $\frac{3}{4} \times \frac{3}{4}$ (f) $1\frac{1}{2} \times 2\frac{1}{2}$ (h) $2\frac{1}{4} \times 1\frac{5}{8}$

2.30. Perform the indicated divisions:

(a) $\dfrac{4/7}{2}$ (c) $\dfrac{7/8}{2}$ (e) $\dfrac{5}{3/4}$ (g) $\dfrac{1\frac{5}{8}}{1\frac{1}{8}}$

(b) $\dfrac{2/3}{4}$ (d) $\dfrac{1/2}{7/8}$ (f) $\dfrac{2}{1\frac{1}{2}}$ (h) $\dfrac{1/4}{2\frac{2}{3}}$

2.31. Find the decimal equivalents of the following numbers:

(a) $1/5$ (b) $4/5$ (c) $1\frac{1}{5}$ (d) $1/6$ (e) $2\frac{3}{4}$ (f) $4\frac{11}{12}$

2.32. Perform the indicated additions and subtractions:

(a) $8 + 4.37$ (b) $6.12 + 5.98$ (c) $30 + 1.578 + 0.02$ (d) $6.3 - 2.1$ (e) $3.5 - 2.9$
(f) $6.1 - 0.031$

2.33. Perform the indicated multiplications and divisions:

(a) 4×2.3 (d) 17×0.01 (g) $1.9/2$ (j) $0.71/0.02$
(b) 7.1×8.4 (e) 0.4×0.02 (h) $0.4/6$
(c) 2.13×4.9 (f) $16/0.4$ (i) $0.48/1.6$

2.34. Change the following percentages to their decimal equivalents:

(a) 41% (b) 17.8% (c) 2% (d) 0.273% (e) 180% (f) 204.3%

2.35. Change the following decimals to their percentage equivalents:

(a) 0.6 (b) 0.633 (c) 0.04 (d) 0.0189 (e) 1.5 (f) 3.485

2.36. Change the following fractions to their percentage equivalents:

(a) $1/4$ (b) $1/8$ (c) $1/16$ (d) $2/5$ (e) $7/32$ (f) $1\frac{7}{8}$

2.37. A discount of 2% is allowed for payment of a bill within 30 days. How large a check is needed to pay a bill of \$147 if this discount is taken advantage of?

2.38. The sales tax on a purchase of \$35 is \$1.75. What is the tax rate?

2.39. A resistor coded as 500 kΩ has an actual resistance of 460 kΩ. What is the percentage of the discrepancy? (1 kΩ = 1000 Ω)

2.40. A voltage of 118 V is measured at the end of a power line known to have a drop of 1.5%. What is the supply voltage?

2.41. A resistor in a certain application is required to dissipate a maximum of 12 W. If a safety margin of 50% is required beyond this, what should the resistor's rating be?

Answers to Supplementary Problems

2.24. (a) 3/5   (b) 1/6   (c) 7/8   (d) 5/12   (e) 1/7   (f) 12/25

2.25. (a) 3   (b) $2\frac{2}{3}$ (c) $2\frac{1}{8}$ (d) $11\frac{7}{8}$ (e) $3\frac{3}{4}$ (f) $2\frac{2}{5}$

2.26. (a) 3/2   (b) 22/3   (c) 29/8   (d) 23/10   (e) 89/6   (f) 31/16

2.27. (a) 1   (c) $1\frac{1}{2}$ (e) $1\frac{43}{60}$ (g) $2\frac{7}{8}$
         (b) $1\frac{1}{2}$ (d) 13/20   (f) 2   (h) $7\frac{1}{6}$

2.28. (a) 1/3   (c) 1/6   (e) $3\frac{3}{4}$ (g) $7\frac{3}{5}$
         (b) 1/4   (d) 11/20   (f) $1\frac{7}{12}$ (h) $-7/8$

2.29. (a) $1\frac{3}{4}$ (c) 1/16   (e) $1\frac{5}{16}$ (g) $3\frac{1}{2}$
         (b) $12\frac{6}{7}$ (d) 9/16   (f) $3\frac{3}{4}$ (h) $3\frac{21}{32}$

2.30. (a) 2/7   (c) 7/16   (e) $6\frac{2}{3}$ (g) $1\frac{4}{9}$
         (b) 1/6   (d) 4/7   (f) $1\frac{1}{3}$ (h) 3/32

2.31. (a) 0.2   (b) 0.8   (c) 1.2   (d) 0.167   (e) 2.75   (f) 4.917

2.32. (a) 12.37   (b) 12.1   (c) 31.598   (d) 4.2   (e) 0.6   (f) 6.069

2.33. (a) 9.2   (c) 10.437   (e) 0.008   (g) 0.95   (i) 0.3
         (b) 59.64   (d) 0.17   (f) 40   (h) 0.0667   (j) 35.5

2.34. (a) 0.41   (b) 0.178   (c) 0.02   (d) 0.00273   (e) 1.8   (f) 2.043

2.35. (a) 60%   (b) 63.3%   (c) 4%   (d) 1.89%   (e) 150%   (f) 348.5%

2.36. (a) 25%   (b) 12.5%   (c) 6.25%   (d) 40%   (e) 21.875%   (f) 1.875%

2.37. $144.06   **2.40.** 119.9 V

2.38. 5%   **2.41.** 18 W

2.39. 8%

Chapter 3

Power and Energy

ELECTRIC POWER

Energy is that property something has which enables it to produce changes (in a general sense) in the physical world. A car moving rapidly has more energy of motion than a car moving slowly; a stone on a high cliff has more energy of position than a stone on the ground; a pail of hot water has more thermal energy than a pail of cold water; the sun emits more radiant energy than a candle; a car battery contains more electric energy than a flashlight battery; and so on. The standard unit of energy is the *joule* (J) although others, such as the foot-pound and the kilowatthour, are often used for particular purposes.

Power (*P*) is the rate at which energy is transported from one place to another or is being changed from one form to another. The more powerful a person, a motor, an amplifier, a heater, or a light bulb is, the greater the amount of energy it can put out in a certain period of time:

$$\text{Power} = \frac{\text{energy}}{\text{time interval}}$$

The unit of power is the *watt* (W), which is equal to 1 J/s. Because the watt is a small unit, the *kilowatt* (kW) is more widely used, where

$$1 \text{ kilowatt} = 1 \text{ kW} = 1000 \text{ watts}$$

The power of an electric current depends both on its magnitude *I* and upon the potential difference *V* across the ends of the circuit. The greater the current and the greater the voltage (which can be thought of as electrical pressure), the greater the power:

$$\text{Electric power} = \text{current} \times \text{voltage}$$

$$P = IV$$

When *I* is in amperes and *V* in volts, *P* will be in watts. This formula can be expressed in the alternative forms

$$\text{Current} = \frac{\text{power}}{\text{voltage}} \qquad \text{Voltage} = \frac{\text{power}}{\text{current}}$$

$$I = \frac{P}{V} \qquad V = \frac{P}{I}$$

Problem 3.1. What is the maximum power in kilowatts that can be carried by a 120-V power line that has a 20-A fuse?

The purpose of a fuse is to limit the current in a circuit to a safe level. The maximum power here is
$$P = IV = 20 \text{ A} \times 120 \text{ V} = 2400 \text{ W} = 2.4 \text{ kW}$$

Problem 3.2. An electric drill rated at 400 W is connected to a 240-V power line. How much current does it draw?
$$I = \frac{P}{V} = \frac{400 \text{ W}}{240 \text{ V}} = 1.67 \text{ A}$$

Problem 3.3. A 10-W resistor is capable of dissipating electric energy as heat at the rate of 10 W. If the current in this resistor is to be 0.04 A in a certain application, what is the highest voltage drop possible across it?

$$V = \frac{P}{I} = \frac{10 \text{ W}}{0.04 \text{ A}} = 250 \text{ V}$$

Problem 3.4. The *horsepower* (hp) is a unit of power equal to $746 \text{ W} = 0.746 \text{ kW}$. A generator driven by a diesel engine that develops 12 hp delivers 30 A at 240 V. What is the efficiency of the generator? (The efficiency of a device is the ratio between the energy or power it delivers and the energy or power it takes in, so that efficiency = output/input.)

Since $1 \text{ hp} = 746 \text{ W}$ and $P = IV$,

$$\text{Input power} = 12 \text{ hp} \times 746 \text{ W/hp} = 8952 \text{ W}$$
$$\text{Output power} = 30 \text{ A} \times 240 \text{ V} = 7200 \text{ W}$$

The efficiency of the generator is therefore

$$\text{Efficiency} = \frac{\text{output}}{\text{input}} = \frac{7200 \text{ W}}{8952 \text{ W}} = 0.80 = 80\%$$

If the conductor or device through which a current passes has the resistance R, the power consumed may be expressed in terms of I and R or in terms of V and R by substituting $V = IR$ or $I = V/R$ in the formula for power $P = IV$. Thus

$$P = IV = I \times IR = I^2 R \qquad P = IV = \frac{V}{R} \times V = \frac{V^2}{R}$$

A review of powers and roots will be useful before applying these expressions to problems.

EXPONENTS

There is a special shorthand way to express a quantity that is to be multiplied by itself one or more times. In this scheme a superscript number called an *exponent* is used to indicate how many times the self-multiplication is to be carried out, as follows:

$$a = a^1$$
$$a \times a = a^2$$
$$a \times a \times a = a^3$$
$$a \times a \times a \times a = a^4$$

and so on.

The quantity a^2 is read as "a squared" because it is equal to the area of a square whose sides are a long, and a^3 is read a "a cubed" because it is equal to the volume of a cube whose edges are a long. Past an exponent of 3 we read a^n as "a to the nth power," so that a^4 is "a to the fourth power."

Problem 3.5. The current in a 50-Ω resistance is 2 A. How much power is dissipated as heat?

$$P = I^2 R = (2 \text{ A})^2 \times 50 \ \Omega = 2 \text{ A} \times 2 \text{ A} \times 50 \ \Omega = 200 \text{ W}$$

Problem 3.6. Find the resistance of a 3000-W, 240-V electric water heater.

The procedure is to solve $P = V^2/R$ for R by transposing the P and the R to give $R = V^2/P$, and then to substitute for V and P:

$$P = \frac{V^2}{R}$$
$$RP = V^2$$
$$R = \frac{V^2}{P} = \frac{(240 \text{ V})^2}{3000 \text{ W}} = \frac{240 \text{ V} \times 240 \text{ V}}{3000 \text{ W}} = 19.2 \ \frac{V^2}{W} = 19.2 \ \Omega$$

Problem 3.7.   The area of a circle whose radius is r is given by the formula $A = \pi r^2$.   Find the area of a circle whose radius is 6 in.

Although $\pi = 3.14159\ldots$,   it is sufficiently accurate here to let $\pi = 3.14$.   Hence

$$A = \pi r^2 = 3.14 \times (6 \text{ in})^2 = 3.14 \times 6 \text{ in} \times 6 \text{ in} = 113 \text{ in}^2$$

The abbreviation "in^2" stands for "square inches."

Problem 3.8.   The volume of a sphere whose radius is r is given by the formula $V = (4/3)\pi r^3$. Find the volume of a spherical tank whose radius is 10 m.

$$V = \frac{4}{3}\pi r^3 = \frac{4}{3} \times 3.14 \times (10 \text{ m})^3 = \frac{4 \times 3.14 \times 10 \text{ m} \times 10 \text{ m} \times 10 \text{ m}}{3} = 4187 \text{ m}^3$$

The abbreviation "m^3" stands for "cubic meters."

ROOTS

When the *square root* of a quantity is multiplied by itself, the product is equal to the quantity. The usual symbol for the square root of a quantity a is $\sqrt{a}$.   Thus

$$\sqrt{a} \times \sqrt{a} = a$$

Problem 3.9.   Examples of square roots.

$\sqrt{1} = 1$	because	$1 \times 1 = 1$
$\sqrt{4} = 2$	because	$2 \times 2 = 4$
$\sqrt{9} = 3$	because	$3 \times 3 = 9$
$\sqrt{100} = 10$	because	$10 \times 10 = 100$
$\sqrt{30.25} = 5.5$	because	$5.5 \times 5.5 = 30.25$
$\sqrt{16B^2} = 4B$	because	$4B \times 4B = 16B^2$

In the case of a number smaller than 1, the square root is larger than the number itself:

$\sqrt{0.01} = 0.1$	because	$0.1 \times 0.1 = 0.01$
$\sqrt{0.25} = 0.5$	because	$0.5 \times 0.5 = 0.25$

Similarly, multiplying the *cube root* $\sqrt[3]{a}$ of a quantity a by itself twice yields the quantity:

$$\sqrt[3]{a} \times \sqrt[3]{a} \times \sqrt[3]{a} = a$$

An expression of the form $\sqrt[n]{a}$ is read as "the nth root of a"; for instance, $\sqrt[4]{16}$ is "the fourth root of 16" and is equal to 2 since $2 \times 2 \times 2 \times 2 = 16$.

Although procedures exist for finding square and cube roots arithmetically, in practice electronic calculators or printed tables are normally used nowadays.   Logarithms (Chapter 4) provide another means to obtain roots.   Tables of roots usually cover whole numbers from 1 to 999.   The tables can be used for the square roots of smaller and larger numbers by keeping in mind these formulas:

For $a = 0.01$–9.99:

$$\sqrt{a} = \frac{\sqrt{100a}}{10}$$

For $a = 1000$–$99\,900$:

$$\sqrt{a} = 10 \times \sqrt{\frac{a}{100}}$$

Problem 3.10. Square roots of small and large numbers.

$$\sqrt{0.07} = \frac{\sqrt{100 \times 0.07}}{10} = \frac{\sqrt{7}}{10} = \frac{2.65}{10} = 0.265$$

$$\sqrt{3.65} = \frac{\sqrt{100 \times 3.65}}{10} = \frac{\sqrt{365}}{10} = \frac{19.1}{10} = 1.91$$

$$\sqrt{4500} = 10 \times \sqrt{\frac{4500}{100}} = 10 \times \sqrt{45} = 10 \times 6.71 = 67.1$$

$$\sqrt{90\,000} = 10 \times \sqrt{\frac{90\,000}{100}} = 10 \times \sqrt{900} = 10 \times 30 = 300$$

What about the square root of a number like 27.8? If we don't have an electronic calculator able to find square roots and don't wish to use logarithms, we can always use a trial-and-error approach. Since $\sqrt{27} = 5.20$ and $\sqrt{28} = 5.29$, we know that $\sqrt{27.8}$ lies somewhere between these numbers. Let us try 5.26: $5.26^2 = 5.26 \times 5.26 = 27.67$. Next we try 5.27: $5.27^2 = 5.27 \times 5.27 = 27.77$, which is pretty close to 27.8. To be sure 5.27 is as close as we can get with three digits, we try 5.28: $5.28^2 = 5.28 \times 5.28 = 27.88$, which is farther away from 27.8 than 27.77. Hence we can say that, to three-digit accuracy, $\sqrt{27.8} = 5.27$.

In Chapter 4 we will see how to use tables for very small and very large numbers without logarithms.

Problem 3.11. The 18-Ω filament of a tube is rated at 2 W. Find its operating voltage.

We first solve $P = V^2/R$ for V^2 by multiplying both sides by R.

$$V^2 = PR$$

Therefore

$$V^2 = PR = 2 \text{ W} \times 18 \text{ Ω} = 36 \text{ V}^2$$

Now we take the square root of both sides of the equation $V^2 = 36 \text{ V}^2$, making use of the fact $\sqrt{36} = 6$:

$$\sqrt{V^2} = V = \sqrt{36 \text{ V}^2} = 6 \text{ V}$$

Problem 3.12. Find the maximum current in a 50-Ω, 10-W resistor if its power rating is not to be exceeded.

Solve $P = I^2R$ for I^2 by dividing both sides by R.

$$I^2 = \frac{P}{R}$$

Therefore

$$I^2 = \frac{P}{R} = \frac{10 \text{ W}}{50 \text{ Ω}} = 0.2 \text{ A}^2$$

Since

$$\sqrt{0.2} = \frac{\sqrt{100 \times 0.2}}{10} = \frac{\sqrt{20}}{10} = \frac{4.47}{10} = 0.447$$

we have

$$\sqrt{I^2} = I = \sqrt{\frac{P}{R}} = \sqrt{0.2 \text{ A}^2} = 0.447 \text{ A}$$

Problem 3.13. Find the radius of a wire whose cross-sectional area is 5 mm². (1 mm = 1 millimeter = 1/1000 meter = 0.0395 in)

As mentioned in Problem 3.7, the area of a circle of radius r is $A = \pi r^2$. Hence

$$A = \pi r^2$$

$$r^2 = \frac{A}{\pi}$$

$$r = \sqrt{\frac{A}{\pi}} = \sqrt{\frac{5 \text{ mm}^2}{3.14}} = \sqrt{1.59 \text{ mm}^2} = 1.26 \text{ mm}$$

ENERGY

The energy W used in a period of time t by a device whose power input is P is given by

$$\text{Energy} = \text{power} \times \text{time interval}$$

$$W = Pt$$

Thus a 400-W electric motor uses in each hour of its operation

$$W = Pt = 400 \text{ W} \times 3600 \text{ s} = 1\,440\,000 \text{ J} = 1.44 \text{ MJ}$$

of energy, since $1 \text{ h} = 60 \text{ min} \times 60 \text{ s/min} = 3600 \text{ s}$. (As mentioned earlier, the *joule* is the standard unit of energy; $1 \text{ MJ} = 1 \text{ megajoule} = 1\,000\,000 \text{ J}$.)

Problem 3.14. The 12-V battery of a certain car has a capacity of 80 ampere-hours (A · h), which means that it can furnish a current of 80 A for 1 h, a current of 40 A for 2 h, and so forth. (a) How much energy is stored in the battery when it is fully charged? (b) If the car's lights require 60 W of power, how long can the battery keep them lit when the engine (and hence its generator) is not running?

(a) The energy the battery can provide is

$$W = Pt = VIt = 12 \text{ V} \times 80 \text{ A} \cdot \text{h} \times 3600 \text{ s/h} = 3\,456\,000 \text{ J} = 3.456 \text{ MJ}$$

(b) Since $W = Pt$, solve for t by dividing both sides by P. Hence

$$t = \frac{W}{P} = \frac{3\,456\,000 \text{ J}}{60 \text{ W}} = 57\,600 \text{ s}$$

To express the answer in hours, we divide by 3600 s/h to give

$$t = \frac{57\,600 \text{ s}}{3600 \text{ s/h}} = 16 \text{ h}$$

Another way to obtain this result is to note that the current drawn by the car's lights is

$$I = \frac{P}{V} = \frac{60 \text{ W}}{12 \text{ V}} = 5 \text{ A}$$

Since the battery's capacity is 80 A · h,

$$t = \frac{80 \text{ A} \cdot \text{h}}{5 \text{ A}} = 16 \text{ h}$$

Perhaps more familiar as a unit of energy than the joule is the *kilowatthour* (kW · h), which is equal to the energy delivered by a source whose power is 1 kilowatt in 1 hour of operation. A 400-W motor uses 0.4 kW · h of energy per hour because $400 \text{ W} = 0.4 \text{ kW}$.

Problem 3.15. A 240-V clothes drier draws a current of 15 A. How much energy in kilowatthours does it use in 45 min of operation?

The power of the drier is

$$P = IV = 15 \text{ A} \times 240 \text{ V} = 3600 \text{ W} = 3.6 \text{ kW}$$

and the time interval here is

$$t = \frac{45 \text{ min}}{60 \text{ min/h}} = 0.75 \text{ h}$$

Hence

$$W = Pt = 3.6 \text{ kW} \times 0.75 \text{ h} = 2.7 \text{ kW} \cdot \text{h}$$

Problem 3.16. A 75-W hall light is left on continuously. At \$0.09/kW · h, find the cost per week.

The power of the light is 0.075 kW and the number of hours in a week is

$$t = 7 \text{ days} \times 24 \text{ h/day} = 168 \text{ h}$$

Hence the energy used per week is

$$W = Pt = 0.075 \text{ kW} \times 168 \text{ h} = 12.6 \text{ kW} \cdot \text{h}$$

and its cost is

$$12.6 \text{ kW} \cdot \text{h} \times \frac{\$0.09}{\text{kW} \cdot \text{h}} = \$1.13$$

Here is a summary of the various formulas for potential difference V, current I, resistance R, and power P that follow from Ohm's law $I = V/R$ and from the power formula $P = VI$.

Unknown Quantity	Known Quantities					
	V, I	I, R	V, R	P, I	P, V	P, R
$V =$		IR		P/I		$\sqrt{PR}$
$I =$			V/R		P/V	$\sqrt{P/R}$
$R =$	V/I			P/I^2	V^2/P	
$P =$	VI	I^2R	V^2/R			

Supplementary Problems

3.17. Find the following squares:
- (a) 5^2
- (b) 12^2
- (c) 30^2
- (d) 400^2
- (e) 1^2
- (f) 0.1^2
- (g) 0.6^2
- (h) 0.03^2

3.18. Find the following cubes:
- (a) 1^3
- (b) 1.2^3
- (c) 3^3
- (d) 10^3
- (e) 12.2^3
- (f) 0.1^3
- (g) 0.3^3
- (h) 0.01^3

3.19. Find the following square roots:
- (a) $\sqrt{16}$
- (b) $\sqrt{121}$
- (c) $\sqrt{42}$
- (d) $\sqrt{0.04}$
- (e) $\sqrt{0.4}$
- (f) $\sqrt{2.5}$
- (g) $\sqrt{625}$
- (h) $\sqrt{36\,000}$

3.20. Find the area of a circle whose radius is 12 cm.

3.21. Find the radius of a circle whose area is 20 ft^2.

3.22. Find the volume of a sphere whose radius is 6 in.

3.23. What is the resistance of a 120-V, 750-W electric iron?

3.24.   What is the current in a 120-V, 100-W light bulb?

3.25.   The current through a 40-Ω resistor is 1.5 A.   How much power is dissipated as heat?

3.26.   A 2-kW heater is to be connected to a 240-V power line whose circuit breaker is rated at 10 A.   What current will the heater draw?   Will the circuit breaker open when the heater is switched on?

3.27.   A light bulb whose power is 100 W when operated at 240 V is instead connected to a 120-V source.   What is the current in the bulb?   How much power does it now dissipate?

3.28.   The starting motor of a car develops 1 hp when it turns the engine over.   How much current does it draw from a 12-V battery, assuming 100% efficiency?

3.29.   What is the power input to an electric motor that draws a current of 4 A when operated at 240 V?   How many horsepower is this?

3.30.   A 1/2-hp electric motor has a power input of 450 W.   Find its percentage efficiency.

3.31.   How much power in kilowatts is needed to drive a 5-kW generator that is 87% efficient?

3.32.   The power output of a certain tube is 8 W.   If the tube is 60% efficient and the plate voltage is 300 V, find the plate current.

3.33.   The power input to a 96%-efficient transformer is 20 kW.   How much current at 240 V does the transformer deliver?   How much power is dissipated as heat?

3.34.   A 12-V storage battery is charged by a current of 20 A at a potential difference of 14 V for 1 h.   How much power is required to charge the battery at this rate?   How much energy has been provided during the process?

3.35.   A 32-V storage battery has a capacity of 1.0 MJ.   For how long can it supply a current of 5 A?

3.36.   The 12-V battery of a car is required to be able to operate the 1.5-kW starting motor for a total of at least 10 min.   What should the minimum capacity of the battery be in ampere-hours?   How much energy is stored in such a battery?

Answers to Supplementary Problems

3.17.	(a) 25	(c) 900	(e) 1	(g) 0.36
	(b) 144	(d) 160 000	(f) 0.01	(h) 0.0009

3.18.	(a) 1	(c) 27	(e) 1815.8	(g) 0.027
	(b) 1.728	(d) 1000	(f) 0.001	(h) 0.000001

3.19.	(a) 4	(c) 6.48	(e) 0.632	(g) 25
	(b) 11	(d) 0.2	(f) 1.58	(h) 189.7

3.20.   452 cm^2 **3.21.**   2.52 ft   **3.22.**   905 in^3

3.23. 19.2 Ω   **3.28.** 62 A   **3.33.** 80 A; 800 W

3.24. 0.83 A   **3.29.** 960 W; 1.3 hp   **3.34.** 280 W; 1.008 MJ

3.25. 90 W   **3.30.** 83%   **3.35.** 6250 s = 1 h 44 min 10 s

3.26. $8\frac{1}{3}$ A; no   **3.31.** 5.75 kW   **3.36.** 21 A · h; 0.9 MJ

3.27. 0.21 A; 25 W   **3.32.** 0.044 A

Chapter 4

Powers of Ten and Logarithms

POWERS OF TEN

A convenient and widely used method of expressing very large and very small numbers makes use of powers of ten. Any number in decimal form can be written as a number between 1 and 10 multiplied by a power of ten, a positive power for numbers larger than 10 and a negative power for numbers smaller than 1. The powers of ten from 10^{-6} to 10^6 are as follows:

$$10^0 \ = 1 \qquad\qquad\quad = 1 \text{ with decimal point moved 0 places}$$
$$10^{-1} = 0.1 \qquad\qquad = 1 \text{ with decimal point moved 1 place to the left}$$
$$10^{-2} = 0.01 \qquad\qquad = 1 \text{ with decimal point moved 2 places to the left}$$
$$10^{-3} = 0.001 \qquad\quad\ = 1 \text{ with decimal point moved 3 places to the left}$$
$$10^{-4} = 0.0001 \qquad\ \ = 1 \text{ with decimal point moved 4 places to the left}$$
$$10^{-5} = 0.000\,01 \ \ = 1 \text{ with decimal point moved 5 places to the left}$$
$$10^{-6} = 0.000\,001 = 1 \text{ with decimal point moved 6 places to the left}$$

$$10^0 \ = 1 \qquad\qquad\quad = 1 \text{ with decimal point moved 0 places}$$
$$10^1 \ = 10 \qquad\qquad\ \ = 1 \text{ with decimal point moved 1 place to the right}$$
$$10^2 \ = 100 \qquad\qquad\ = 1 \text{ with decimal point moved 2 places to the right}$$
$$10^3 \ = 1000 \qquad\qquad = 1 \text{ with decimal point moved 3 places to the right}$$
$$10^4 \ = 10\,000 \qquad\ \ = 1 \text{ with decimal point moved 4 places to the right}$$
$$10^5 \ = 100\,000 \qquad = 1 \text{ with decimal point moved 5 places to the right}$$
$$10^6 \ = 1\,000\,000 = 1 \text{ with decimal point moved 6 places to the right}$$

Problem 4.1. Examples of powers-of-ten notation.

$$20 = 2 \times 10 = 2 \times 10^2 \qquad\qquad 0.22 = 2.2 \times 0.1 = 2.2 \times 10^{-1}$$
$$3043 = 3.043 \times 1000 = 3.043 \times 10^3 \qquad 0.000\,035 = 3.5 \times 0.000\,01 = 3.5 \times 10^{-5}$$
$$8\,700\,000 = 8.7 \times 1\,000\,000 = 8.7 \times 10^6$$

When numbers written in powers-of-10 notation are to be added or subtracted, they must all be expressed in terms of the *same* power of 10:

$$3 \times 10^2 + 4 \times 10^3 = 0.3 \times 10^3 + 4 \times 10^3 = 4.3 \times 10^3$$

It does not matter which power of 10 is used, as long as it is the same one for all the numbers. Thus we get the same answer as before if we write instead

$$3 \times 10^2 + 4 \times 10^3 = 3 \times 10^2 + 40 \times 10^2 = 43 \times 10^2 = 4.3 \times 10^3$$

Since a step is saved if the power of 10 used is that of the larger number, it makes sense to do this, but it is purely for the sake of convenience.

Problem 4.2. Examples of addition.

$$6 \times 10^2 + 5 \times 10^4 = 0.06 \times 10^4 + 5 \times 10^4 = 5.06 \times 10^4 \qquad 7 + 2 \times 10^2 = 7 + 200 = 207$$
$$2 \times 10^{-2} + 3 \times 10^{-3} = 2 \times 10^{-2} + 0.3 \times 10^{-2} = 2.3 \times 10^{-2} \qquad 7 + 2 \times 10^{-2} = 7 + 0.02 = 7.02$$

Note that 10^{-2} is larger than 10^{-3}.

30

To subtract one number from another, the same procedure is followed. If the number being subtracted is the larger of the two, the answer will be negative in sign, just as $3 - 5 = -2$.

Problem 4.3.　Examples of subtraction.

$$6 \times 10^4 - 4 \times 10^2 = 6 \times 10^4 - 0.04 \times 10^4 = 5.96 \times 10^4$$
$$3 \times 10^{-2} - 5 \times 10^{-3} = 3 \times 10^{-2} - 0.5 \times 10^{-2} = 2.5 \times 10^{-2}$$
$$9 - 2 \times 10^{-3} = 9 - 0.002 = 8.998$$
$$7 \times 10^{-5} - 2 \times 10^{-4} = 0.7 \times 10^{-4} - 2 \times 10^{-4} = -1.3 \times 10^{-4}$$
$$2 \times 10^3 - 5 \times 10^5 = 0.02 \times 10^5 - 5 \times 10^5 = -4.98 \times 10^5$$
$$4 - 3 \times 10^2 = 0.04 \times 10^2 - 3 \times 10^2 = -2.96 \times 10^2$$

In subtraction, the answer must sometimes be expressed using a smaller power of 10.

Problem 4.4.　Additional examples of subtraction.

$$3.34 \times 10^3 - 3.20 \times 10^3 = 0.14 \times 10^3 = 1.4 \times 10^2$$
$$5.47 \times 10^{-5} - 5.43 \times 10^{-5} = 0.04 \times 10^{-5} = 4 \times 10^{-7}$$
$$7.23 \times 10^{-3} - 7.28 \times 10^{-3} = -0.05 \times 10^{-3} = -5 \times 10^{-5}$$

MULTIPLICATION AND DIVISION

An advantage of powers-of-10 notation is that it makes calculations that involve large and small numbers relatively easy to carry out. To multiply two powers of 10, the procedure is just to add their exponents:

$$10^n \times 10^m = 10^{n+m}$$

Let us work out $10^2 \times 10^3$ to verify this rule. Since

$$10^2 = 10 \times 10 \qquad \text{and} \qquad 10^3 = 10 \times 10 \times 10$$

we have

$$10^2 \times 10^3 = (10 \times 10) \times (10 \times 10 \times 10) = 10 \times 10 \times 10 \times 10 \times 10 = 10^5$$

Problem 4.5.　Examples of multiplication.

$$4 \times 10^2 \times 10^6 = 4 \times 10^8 \qquad\qquad 5 \times 10^{-4} \times 6 \times 10^{-4} = 30 \times 10^{-8} = 3 \times 10^{-7}$$
$$7 \times 10^2 \times 5 \times 10^3 = 35 \times 10^5 = 3.5 \times 10^6 \qquad\qquad 10^5 \times 10^{-2} = 10^{5-2} = 10^3$$
$$2 \times 10^{-4} \times 4 \times 10^{-4} = 8 \times 10^{-8} \qquad\qquad 7 \times 10^{-5} \times 3 \times 10^5 = 21 \times 10^0 = 21 \times 1 = 21$$

To divide a power of 10 by another power of 10, subtract the exponent of the second power of 10 from the exponent of the first:

$$\frac{10^n}{10^m} = 10^{n-m}$$

For example,

$$\frac{10^5}{10^3} = 10^{5-3} = 10^2$$

Let us work out $10^5/10^3$ to verify this rule. Since

$$10^5 = 10 \times 10 \times 10 \times 10 \times 10 \qquad \text{and} \qquad 10^3 = 10 \times 10 \times 10$$

we have

$$\frac{10^5}{10^3} = \frac{10 \times 10 \times 10 \times 10 \times 10}{10 \times 10 \times 10} = 10 \times 10 = 10^2$$

Problem 4.6. Examples of division.

$$\frac{10^3}{10^7} = 10^{3-7} = 10^{-4}$$

$$\frac{3 \times 10^6}{10^4} = 3 \times 10^2$$

$$\frac{10^4}{8 \times 10^2} = \frac{1 \times 10^4}{8 \times 10^2} = \frac{1}{8} \times \frac{10^4}{10^2} = 0.125 \times 10^2 = 12.5$$

$$\frac{8 \times 10^6}{2 \times 10^3} = \frac{8}{2} \times \frac{10^6}{10^3} = 4 \times 10^3$$

$$\frac{10^2}{8} = \frac{1 \times 10^2}{8} = \frac{1}{8} \times 10^2 = 0.125 \times 10^2 = 0.125 \times 100 = 12.5$$

$$\frac{8 \times 10^3}{2 \times 10^6} = \frac{8}{2} \times \frac{10^3}{10^6} = 4 \times 10^{-3}$$

$$\frac{8 \times 10^3}{2 \times 10^3} = \frac{8}{2} \times \frac{10^3}{10^3} = 4$$

$$\frac{5 \times 10^4}{2 \times 10^{-2}} = \frac{5}{2} \times \frac{10^4}{10^{-2}} = 2.5 \times 10^{4-(-2)} = 2.5 \times 10^{4+2} = 2.5 \times 10^6$$

Reciprocals follow the pattern

$$\frac{1}{10^n} = 10^{-n}$$

For example,

$$\frac{1}{10^3} = 10^{-3}$$

To verify this rule, we note that

$$\frac{1}{10^3} = \frac{1}{10 \times 10 \times 10} = \frac{1}{1000} = 0.001 = 10^{-3}$$

Problem 4.7. Examples of reciprocals.

$$\frac{1}{10^{-5}} = 10^{-(-5)} = 10^5$$

$$\frac{1}{2 \times 10^6} = \frac{1 \times 1}{2 \times 10^6} = \frac{1}{2} \times \frac{1}{10^6} = 0.5 \times 10^{-6} = 5 \times 10^{-7}$$

$$\frac{1}{4 \times 10^{-3}} = \frac{1 \times 1}{4 \times 10^{-3}} = \frac{1}{4} \times \frac{1}{10^{-3}} = 0.25 \times 10^3 = 2.5 \times 10^2$$

Problem 4.8. A sample calculation.

$$\frac{460 \times 0.000\,03 \times 100\,000}{9000 \times 0.0062} = \frac{(4.6 \times 10^2) \times (3 \times 10^{-5}) \times (10^5)}{(9 \times 10^3) \times (6.2 \times 10^{-3})} = \frac{4.6 \times 3}{9 \times 6.2} \times \frac{10^2 \times 10^{-5} \times 10^5}{10^3 \times 10^{-3}}$$

$$= 0.25 \times \frac{10^{2-5+5}}{10^{3-3}} = 0.25 \times \frac{10^2}{10^0} = 25$$

POWERS AND ROOTS

Raising a power of ten to a power involves multiplying the two exponents together to find the new exponent.   In symbols,

$$(10^n)^m = 10^{n \times m}$$

Problem 4.9.   Examples of powers of numbers.

$(10^4)^2 = 10^{4 \times 2} = 10^8$ $\qquad\qquad$ $(3 \times 10^3)^2 = 3^2 \times (10^3)^2 = 9 \times 10^6$

$(10^2)^4 = 10^{2 \times 4} = 10^8$ $\qquad\qquad$ $(4 \times 10^{-5})^3 = 4^3 \times (10^{-5})^3 = 64 \times 10^{-15} = 6.4 \times 10^{-14}$

$(10^{-3})^2 = 10^{-3 \times 2} = 10^{-6}$ $\qquad$ $(2 \times 10^{-2})^{-4} = \dfrac{1}{2^4} \times (10^{-2})^{-4} = \dfrac{1}{16} \times 10^8 = 0.0625 \times 10^8 = 6.25 \times 10^6$

$(10^{-4})^{-3} = 10^{-4 \times (-3)} = 10^{12}$

To find the square root of a number expressed in powers-of-ten notation, the power of ten must be an even number.   The square root of the power of ten is then found by dividing the exponent by 2. In symbols,

$$\sqrt{10^n} = 10^{n/2}$$

Problem 4.10.   Examples of square roots involving even exponents.

$\sqrt{10^6} = 10^{6/2} = 10^3$ $\qquad\qquad$ $\sqrt{1.3 \times 10^{-8}} = \sqrt{1.3} \times \sqrt{10^{-8}} = 1.14 \times 10^{-4}$

$\sqrt{10^{-4}} = 10^{-4/2} = 10^{-2}$ $\qquad\qquad$ $\sqrt{9 \times 10^{-2}} = \sqrt{9} \times \sqrt{10^{-2}} = 3 \times 10^{-1} = 0.3$

$\sqrt{5 \times 10^4} = \sqrt{5} \times \sqrt{10^4} = 2.24 \times 10^2$

When the power of ten is odd, the rule is to change it to the next smaller power while multiplying the number in front by 10.   This gives an even power of ten and the resulting square root will be in the normal form.

Problem 4.11.   Examples of square roots involving odd exponents.

$$\sqrt{10^5} = \sqrt{1 \times 10^5} = \sqrt{10 \times 10^4} = \sqrt{10} \times \sqrt{10^4} = 3.16 \times 10^2$$

$$\sqrt{10^{-3}} = \sqrt{1 \times 10^{-3}} = \sqrt{10 \times 10^{-4}} = \sqrt{10} \times \sqrt{10^{-4}} = 3.16 \times 10^{-2}$$

$$\sqrt{3 \times 10^5} = \sqrt{30 \times 10^4} = \sqrt{30} \times \sqrt{10^4} = 5.48 \times 10^2$$

$$\sqrt{6 \times 10^{-7}} = \sqrt{60 \times 10^{-8}} = \sqrt{60} \times \sqrt{10^{-8}} = 7.75 \times 10^{-4}$$

$$\sqrt{0.000\,025} = \sqrt{2.5 \times 10^{-5}} = \sqrt{25 \times 10^{-6}} = \sqrt{25} \times \sqrt{10^{-6}} = 5 \times 10^{-3}$$

UNITS

Units are algebraic quantities and may be multiplied and divided by one another.   To convert a quantity expressed in a certain unit to its equivalent in a different unit of the same kind, we use the fact that multiplying or dividing anything by 1 does not affect its value.   For instance: 12 in = 1 ft, so   12 in/ft = 1,   and we can convert a length s expressed in feet to its value in inches by multiplying s by 12 in/ft:

$$4 \text{ ft} = 4 \,\cancel{\text{ft}} \times 12 \,\frac{\text{in}}{\cancel{\text{ft}}} = 48 \text{ in}$$

Conversion factors for the most common U.S. Customary and SI (metric) units are given in Appendix A.

Problem 4.12.   Rome is 1400 km by road from Paris.   How far is this in miles?

Since   1 km = 0.621 mi,

$$1440 \text{ km} = 1440 \text{ k\!m} \times 0.621 \frac{\text{mi}}{\text{k\!m}} = 894 \text{ mi}$$

Problem 4.13.   A man is 6 ft 2 in tall.   How many centimeters is this?

Since   1 ft = 12 in   and   1 in = 2.54 cm,

$$6 \text{ ft } 2 \text{ in} = [(6 \times 12) + 2] \text{ in} = 74 \text{ i\!n} \times 2.54 \frac{\text{cm}}{\text{i\!n}} = 188 \text{ cm}$$

Problem 4.14.   How many square feet are there in 1 m²?

Since   1 m = 3.28 ft,

$$1 \text{ m}^2 = 1 \text{ m}^2 \times \left(3.28 \frac{\text{ft}}{\text{m}} \right)^2 = 10.76 \text{ ft}^2$$

Problem 4.15.   Express a velocity of 60 mi/h in feet per second.

There are 5280 ft in a mile and 3600 s in an hour, and so

$$60 \frac{\text{mi}}{\text{h}} = 60 \frac{\text{m\!i}}{\text{h}} \times 5280 \frac{\text{ft}}{\text{m\!i}} \times \frac{1}{3600 \text{ s} / \text{h}} = 88 \frac{\text{ft}}{\text{s}}$$

Subdivisions and multiples of metric units are designated by prefixes according to the corresponding power of 10.

Prefix	Power	Abbreviation	Example
pico-	10^{-12}	p	1 pf = 1 picofarad = 10^{-12} farad
nano-	10^{-9}	n	1 ns = 1 nanosecond = 10^{-9} second
micro-	10^{-6}	μ	1 μA = 1 microampere = 10^{-6} ampere
milli-	10^{-3}	m	1 mm = 1 millimeter = 10^{-3} meter
centi-	10^{-2}	c	1 cl = 1 centiliter = 10^{-2} liter
kilo-	10^{3}	k	1 kg = 1 kilogram = 10^{3} grams
mega-	10^{6}	M	1 MW = 1 megawatt = 10^{6} watts
giga-	10^{9}	G	1 GeV = 1 gigaelectronvolt = 10^{9} electronvolts

Problem 4.16.   When the voltage across a capacitor of capacitance C is V, the charge on the capacitor is $Q = CV$.   Find the charge on a 200-pF capacitor when it is connected to a 3-kV source.

Here $C = 200 \text{ pF} = 200 \text{ picofarads} = 200 \times 10^{-12} \text{ F} = 2 \times 10^{-10} \text{ F}$ and $V = 3 \text{ kV} = 3 \text{ kilovolts} = 3 \times 10^3 \text{ V}$,   and so

$$Q = CV = 2 \times 10^{-10} \text{ F} \times 3 \times 10^3 \text{ V} = 6 \times 10^{-7} \text{ C}$$

The answer can be expressed as $Q = 0.6 \ \mu\text{C} = 0.6$ microcoulombs since $1 \ \mu\text{C} = 10^{-6}$ C.

Problem 4.17.   The inductive reactance of a coil of inductance L when the frequency of the current in it is f is given by $X_L = 2\pi fL$.   Find the inductive reactance of a 5-mH inductor when the current in it has the frequency 10 kHz.

Here $L = 5 \text{ mH} = 5 \text{ millihenries} = 5 \times 10^{-3} \text{ H}$ and $f = 10 \text{ kHz} = 10 \text{ kilohertz} = 10 \times 10^3 \text{ Hz} = 10^4 \text{ Hz}$, and so

$$X_L = 2\pi fL = 2\pi \times 10^4 \text{ Hz} \times 5 \times 10^{-3} \text{ H} = 2\pi \times 5 \times 10^{4-3} \ \Omega = 31.4 \times 10^1 \ \Omega = 314 \ \Omega$$

Problem 4.18.   The capacitive reactance of a capacitor of capacitance C when the frequency of the potential difference across it is f is given by $X_C = 1/(2\pi fC)$.   Find the capacitive reactance of a 2-μF capacitor when it is connected to a 0.1-MHz source.

Here $C = 2\ \mu F = 2$ microfarads $= 2 \times 10^{-6}$ F   and $f = 0.1$ MHz $= 0.1$ megahertz $= 0.1 \times 10^6$ Hz,   and so

$$X_C = \frac{1}{2\pi f C} = \frac{1}{2\pi \times 0.1 \times 10^6 \text{ Hz} \times 2 \times 10^{-6} \text{ F}} = \frac{1}{0.4\pi \times 10^{6-6}} \Omega = \frac{1}{0.4\pi \times 10^0} \Omega$$

Since $10^0 = 1$,

$$X_C = \frac{1}{0.4\pi} \Omega = 0.80\ \Omega$$

Problem 4.19.   The resonant frequency of an alternating-current circuit that contains an inductance L and a capacitance C in series is given by $f_0 = 1/(2\pi\sqrt{LC}\,)$.   Find the resonant frequency of a circuit in which $L = 5$ mH   and $C = 50\ \mu F$.

Here $L = 5$ mH $= 5 \times 10^{-3}$ H   and $C = 50\ \mu F = 50 \times 10^{-6}$ F,   and so

$$f_0 = \frac{1}{2\pi\sqrt{LC}} = \frac{1}{2\pi\sqrt{5 \times 10^{-3} \text{ H} \times 50 \times 10^{-6} \text{ F}}} = \frac{1}{2\pi\sqrt{5 \times 50 \times 10^{-3-6}}} \text{ Hz} = \frac{1}{2\pi\sqrt{250 \times 10^{-9}}} \text{ Hz}$$

To find the square root, we change 250×10^{-9} to 25×10^{-8}, with the result

$$f_0 = \frac{1}{2\pi\sqrt{25 \times 10^{-8}}} \text{ Hz} = \frac{1}{2\pi\sqrt{25} \times \sqrt{10^{-8}}} \text{ Hz} = \frac{1}{2\pi \times 5 \times 10^{-4}} \text{ Hz} = \frac{1}{10\pi} \times \frac{1}{10^{-4}} \text{ Hz}$$

$$= 0.0318 \times 10^4 \text{ Hz} = 318 \text{ Hz}$$

With practice, most of the above steps will not have to be written down when working out a calculation of this kind.

LOGARITHMS

Although logarithms have many other uses, their chief application in basic electricity and electronics is in connection with the decibel. Hence logarithms will be discussed here only to the extent required for this purpose.

The *logarithm* of a number N is the power n to which 10 must be raised in order that $10^n = N$. That is,

$$N = 10^n \qquad \text{therefore} \qquad \log N = n$$

(Logarithms are not limited to a base of 10, but base 10 logarithms are the most common and are all that are needed here.)   For instance,

$$1000 = 10^3 \qquad \text{therefore} \qquad \log 1000 = 3$$

$$0.01 = 10^{-2} \qquad \text{therefore} \qquad \log 0.01 = -2$$

Logarithms are not limited to powers of ten that are whole numbers.   For instance,

$$5 = 10^{0.699} \qquad \text{therefore} \qquad \log 5 = 0.699$$

$$240 = 10^{2.380} \qquad \text{therefore} \qquad \log 240 = 2.380$$

Logarithms are only defined for positive numbers: the quantity 10^n is positive whether n is negative, positive, or 0, and since n is the logarithm of 10^n, it can only describe a positive number.

While it is naturally easiest to find the logarithm of a number with an electronic calculator, a table of logarithms is not hard to use.   Such a table usually contains the logarithms of numbers from 1.00 to 9.99.   Logarithms of such numbers range from 0 to 0.9999.   From this table we can immediately find the logarithm of a number in this range, or, given a logarithm, we can find the corresponding number (which is called the *antilogarithm*).   As an example, here is where the logarithm of the number 5.73 is located in the table:

N	0	1	2	3	4
55	7404	7412	7419	7427	7435
56	7482	7490	7497	7505	7513
57	7559	7566	7574	(7582)	7589
58	7634	7642	7649	7657	7664
59	7709	7716	7723	7731	7738

Hence

$$\log 5.73 = 0.7582$$

To find the antilogarithm of a logarithm, we look through the body of the table for the logarithm whose value is closest to the one we have in mind.   If we are given that

$$\log N = 0.7497$$

we see from the table that

$$N = 5.62$$

since $\log 5.62 = 0.7497$. If the given logarithm is not the same as one of those in the table, the corresponding number lies between two adjacent numbers for which logarithms are given.   For instance, the number whose logarithm is 0.7711 is greater than 5.90 but smaller than 5.91 since

$$\log 5.90 = 0.7709$$
$$\log 5.91 = 0.7716$$

Since 0.7711 is closer to $\log 5.90$ than to $\log 5.91$, we are justified in saying that

$$\log N = 0.7711 \qquad \text{therefore} \qquad N = 5.90$$

LOGARITHMS OF LARGE NUMBERS

What do we do if we need the logarithm of a number larger than 10? (The procedure for numbers smaller than 1 will not be considered here since it is more complicated and not needed for use with decibels.)   Because of the way they are defined,

$$\log xy = \log x + \log y$$

Thus the logarithm of any number written in powers-of-ten notation can be expressed as the sum of two logarithms, one of which we can find in a table and the other of which is simply the power of ten involved.   For instance, since

$$804 = 8.04 \times 10^2$$

we have

$$\log 804 = \log 8.04 + \log 10^2 = 0.9053 + 2 = 2.9053$$

The part of the logarithm after the decimal point, here 9053, describes the *sequence of digits* in the original number.   The other part of the logarithm, here 2, describes the *magnitude* of the original number, that is, whether it is 8.04, 80.4, 8040, or, as in this case, 804.

Similarly, to find the number whose logarithm is, say, 3.8751, we begin by separating the logarithm into two parts, one of which we use to determine the digits in the number and the other to locate the decimal point:

$$\text{antilog } 3.8751 = \text{antilog } 3 \times \text{antilog } 0.8751 = 10^3 \times 7.500 = 7500$$

To find an antilogarithm with a calculator, enter the value of the logarithm and press the 10^x key (INV LOG on some calculators).

Problem 4.20.   Examples of logarithms.

$$\log 1 = \log 10^0 = 0$$
$$\log 6 = 0.7782$$
$$\log 10 = \log 10^1 = 1$$
$$\log 27 = \log(2.7 \times 10^1) = \log 2.7 + \log 10 = 0.4314 + 1 = 1.4314$$
$$\log 100 = \log 10^2 = 2$$
$$\log 1000 = \log 10^3 = 3$$
$$\log 10\,000 = \log 10^4 = 4$$
$$\log 90\,200 = \log(9.02 \times 10^4) = \log 9.02 + \log 10^4 = 0.9552 + 4 = 4.9552$$

Problem 4.21. Examples of antilogarithms.

$$\text{antilog } 0 = 10^0 = 1$$
$$\text{antilog } 0.6435 = 4.40$$
$$\text{antilog } 1 = 10^1 = 10$$
$$\text{antilog } 1.4 = \text{antilog } 1 \times \text{antilog } 0.4 = 10^1 \times 2.51 = 25.1$$
$$\text{antilog } 2 = 10^2 = 100$$
$$\text{antilog } 3 = 10^3 = 1000$$
$$\text{antilog } 3.93 = \text{antilog } 3 \times \text{antilog } 0.93 = 1000 \times 8.51 = 8510$$

DECIBELS

The human ear does not respond linearly to sound intensity. If the intensity of a particular sound is doubled, the new sound seems much less than twice as loud. For this reason power ratios in electronics are commonly expressed in terms of a logarithmic unit called the *decibel* (dB); the use of the decibel is so established that it is used even for nonaudio applications.

If the power input to an amplifier or other signal processing device is P_{in} and the power output of the device is P_{out}, the *power gain G* of the system in decibels is defined as

$$G \text{ (dB)} = 10 \log \frac{P_{out}}{P_{in}}$$

A change in audio power output of 1 dB is about the minimum that can be detected by a person with good hearing; usually the change must be 2 or 3 dB to be apparent.

Problem 4.22. Find the power gain of an amplifier whose power input in 0.2 W and whose power output is 80 W.

$$G \text{ (dB)} = 10 \log \frac{P_{out}}{P_{in}} = 10 \log \frac{80 \text{ W}}{0.2 \text{ W}} = 10 \log 400 = 10 \log(4 \times 10^2)$$
$$= 10(\log 4 + \log 10^2) = 10(0.60 + 2) = 10(2.60) = 26 \text{ dB}$$

Of course, with a calculator the value of log 400 can be found in a single step.

Problem 4.23. An input power of 10^{-9} W from a microphone to an amplifier results in a current of 1.5 A in the 8 Ω voice coil of a loudspeaker. What is the power gain of the amplifier?

The output power is

$$P_{out} = I^2 R = (1.5 \text{ A})^2 \times 8 \text{ }\Omega = 18 \text{ W}$$

and so

$$G \text{ (dB)} = 10 \log \frac{P_{out}}{P_{in}} = 10 \log \frac{18 \text{ W}}{10^{-9} \text{ W}} = 10 \log(18 \times 10^9)$$
$$= 10 \log(1.8 \times 10^{10}) = 10(\log 1.8 + \log 10^{10})$$
$$= 10(0.255 + 10) = 10(10.255) = 103 \text{ dB}$$

Problem 4.24. An amplifier has a power gain of 40 dB. What is the actual ratio between its output and input powers?

We start with the definition

$$G \text{ (dB)} = 10 \log \frac{P_{out}}{P_{in}}$$

and divide both sides by 10 to give

$$\frac{G \text{ (dB)}}{10} = \log \frac{P_{out}}{P_{in}}$$

Now we take the antilogarithm of both sides:

$$\text{antilog}\,\frac{G\,(\text{dB})}{10} = \frac{P_{\text{out}}}{P_{\text{in}}}$$

Here $G\,(\text{dB}) = 40$, so the required power ratio is

$$\frac{P_{\text{out}}}{P_{\text{in}}} = \text{antilog}\,\frac{40}{10} = \text{antilog}\,4 = 10^4 = 10\,000$$

Problem 4.25. A record player pickup has an output of 0.002 μW. What is the output power when a 100-dB amplifier is used with it?

From Problem 4.24,

$$\text{antilog}\,\frac{G\,(\text{dB})}{10} = \frac{P_{\text{out}}}{P_{\text{in}}}$$

Hence

$$P_{\text{out}} = P_{\text{in}}\,\text{antilog}\,\frac{G\,(\text{dB})}{10} = 2 \times 10^{-9}\,\text{W} \times \text{antilog}\,\frac{100}{10}$$

$$= 2 \times 10^{-9}\,\text{W} \times \text{antilog}\,10 = 2 \times 10^{-9}\,\text{W} \times 10^{10} = 2 \times 10^{1}\,\text{W} = 20\,\text{W}$$

POWER LOSS

When the output power from a device is less than the input power, the usual procedure is to invert the power ratio and use a minus sign with the logarithm:

$$\log\,\frac{P_{\text{out}}}{P_{\text{in}}} = -\log\,\frac{P_{\text{in}}}{P_{\text{out}}}$$

This gives a negative power gain, which corresponds to a smaller output power. Hence we have

$$G\,(\text{dB}) = -10\log\,\frac{P_{\text{in}}}{P_{\text{out}}}$$

and

$$\frac{P_{\text{in}}}{P_{\text{out}}} = \text{antilog}\,\frac{-G\,(\text{dB})}{10}$$

Problem 4.26. An RG-58/U coaxial cable has a signal attenuation of 7 dB per 100 ft at a frequency of 160 mHz. A 50-ft length of this cable is used to couple a 25-W VHF radio transmitter that operates at 160 mHz to an antenna. How much power reaches the antenna?

"Attenuation" refers to a loss of power, so the power gain of this type of cable is -7 dB per 100 ft. Since the cable length here is 50 ft, the power gain is -7 dB/2 $= -3.5$ dB, and

$$\frac{P_{\text{in}}}{P_{\text{out}}} = \text{antilog}\,\frac{-G\,(\text{dB})}{10} = \text{antilog}\,\frac{-(-3.5)}{10} = \text{antilog}\,0.35 = 2.24$$

$$P_{\text{out}} = \frac{P_{\text{in}}}{2.24} = \frac{25\,\text{W}}{2.24} = 11.2\,\text{W}$$

Less than half the power reaches the antenna.

REFERENCE POWER LEVEL

A certain power level is often used as a standard value. For example, a common reference power level in audio work is 1 mW = 1 milliwatt = 0.001 W = 10^{-3} W. A power output G_{out} can be expressed in terms of this level in decibels by letting $P_{\text{in}} = 1$ mW. To avoid confusion, it is wise to designate such powers by using dBm (m for milliwatt) instead of just dB, although this is often not done.   Then

$$G_{out}\,(dBm) = 10 \log \frac{P_{out}}{1\ mW}$$

If the output power P_{out} is less than the reference level, then

$$G_{out}\,(dBm) = -10 \log \frac{1\ mW}{P_{out}}$$

The unit dBm is sometimes called a *volume unit*, abbreviated VU.

Problem 4.27.   A crystal microphone has an output of 10^{-9} W.   What is its output in dBm?

$$G_{out}\,(dBm) = -10 \log \frac{10^{-3}\ W}{10^{-9}\ W} = -10 \log 10^6 = -10 \times 6 = -60\ dBm$$

OVERALL GAIN

As mentioned earlier, the logarithm of a product is equal to the sum of the logarithms of the factors:

$$\log xy = \log x + \log y$$

This holds true for any number of factors.   For instance, in the case of three factors

$$\log xyz = \log x + \log y + \log z$$

Since power gains in decibels are logarithmic quantities, the overall gain in decibels of a system of several devices is equal to the sum of the separate gains in decibels of the devices:

$$G\,(overall) = G_1\,(dB) + G_2\,(dB) + G_3\,(dB) + \cdots$$

Problem 4.28.   An audio system is made up of components with the following power gains: preamplifier, $+35$ dB; attenuator, -10 dB; amplifier, $+70$ dB.   What is the overall gain of the system?

$$G\,(overall) = +35\ dB - 10\ dB + 70\ dB = +95\ dB$$

Problem 4.29.   A microphone with an ouput of -60 dBm is connected to the audio system of Problem 4.28.   (*a*) What is the system output in dBm?   (*b*) What is the system output in watts?

(*a*)   When a gain in dBm is added to a gain in dB, the result will be in dBm.   The system output here is therefore

$$G\,(overall) = G_1 + G_2 = -60\ dBm + 95\ dB = +35\ dBm$$

(*b*)   Since

$$G\,(dBm) = 10 \log \frac{P_{out}}{1\ mW}$$

we have here

$$\frac{P_{out}}{1\ mW} = antilog\ \frac{G\,(dBm)}{10} = antilog\ \frac{35}{10} = antilog\ 3.5$$

$$= antilog\ 3 \times antilog\ 0.5 = 10^3 \times 3.162 = 3162$$

Hence

$$P_{out} = 1\ mW \times 3162 = 3162\ mW = 3.162\ W$$

since 1 mW = 0.001 W.

Supplementary Problems

4.30. Express the following numbers in power-of-10 notation:

(a) 720   (d) 0.000 062   (g) 49 527   (j) 49 000 000 000

(b) 890 000   (e) 3.6   (h) 0.002 943   (k) 0.000 000 011

(c) 0.02   (f) 0.4   (i) 0.0014   (l) 1.4763

4.31. Express the following numbers in decimal notation:

(a) 3×10^{-4} (c) 8.126×10^{-5} (e) 5×10^2 (g) 4.32145×10^3 (i) 5.7×10^0

(b) 7.5×10^3 (d) 1.01×10^8 (f) 3.2×10^{-2} (h) 6×10^6 (j) 6.9×10^{-5}

4.32. Perform the following additions and subtractions:

(a) $3 \times 10^2 + 4 \times 10^3$ (f) $6.32 \times 10^2 + 5$ (k) $4.6 \times 10^5 - 3.2 \times 10^7$

(b) $2 \times 10^4 + 5 \times 10^6$ (g) $4 \times 10^3 - 3 \times 10^2$ (l) $3 \times 10^5 - 2.98 \times 10^5$

(c) $7 \times 10^{-2} + 2 \times 10^{-3}$ (h) $5 \times 10^7 - 9 \times 10^4$ (m) $4.76 \times 10^{-3} - 4.81 \times 10^{-3}$

(d) $4 \times 10^{-5} + 5 \times 10^{-3}$ (i) $3.2 \times 10^{-4} - 5 \times 10^{-5}$ (n) $7 \times 10^3 + 5 \times 10^2 - 9 \times 10^2$

(e) $2 \times 10^1 + 2 \times 10^{-1}$ (j) $7 \times 10^4 - 2 \times 10^5$ (o) $3 \times 10^{-4} + 6 \times 10^{-5} - 7 \times 10^{-3}$

4.33. Evaluate the following reciprocals:

(a) $\dfrac{1}{10^2}$ (c) $\dfrac{1}{6 \times 10^3}$.   (e) $\dfrac{1}{2 \times 10^{-2}}$

(b) $\dfrac{1}{2 \times 10^2}$ (d) $\dfrac{1}{10^{-2}}$ (f) $\dfrac{1}{4 \times 10^{-4}}$

4.34. Perform the following multiplications and divisions using powers-of-10 notation:

(a) 5000×0.005 (e) $\dfrac{30\,000 \times 0.000\,000\,6}{1000 \times 0.02}$ (h) $\dfrac{0.002 \times 0.000\,000\,05}{0.000\,004}$

(b) $\dfrac{5000}{0.005}$ (f) $\dfrac{0.0001}{60\,000 \times 200}$ (i) $\dfrac{400 \times 0.000\,06}{0.2 \times 20\,000}$

(c) $\dfrac{500\,000 \times 18\,000}{9\,000\,000}$ (g) $\dfrac{200 \times 0.000\,04}{400\,000}$ (j) $\dfrac{0.06 \times 0.0001}{0.000\,03 \times 40\,000}$

(d) $\dfrac{30 \times 80\,000\,000\,000}{0.0004}$

4.35. Evaluate the following and express the results in powers-of-10 notation:

(a) $(4 \times 10^9)^3$ (c) $(2 \times 10^7)^{-2}$ (e) $(3 \times 10^{-8})^2$ (g) $(3 \times 10^{-4})^{-3}$

(b) $(2 \times 10^7)^2$ (d) $(2 \times 10^{-2})^5$ (f) $(5 \times 10^{11})^{-2}$

4.36. Evaluate the following and express the results in powers-of-10 notation. Note that $\sqrt{4} = 2$, $\sqrt{40} = 6.3$, $\sqrt[3]{4} = 1.6$, $\sqrt[3]{40} = 3.4$, and $\sqrt[3]{400} = 7.4$.

(a) $\sqrt{4 \times 10^6}$ (e) $\sqrt{4 \times 10^{-5}}$ (h) $\sqrt[3]{4 \times 10^{14}}$ (k) $\sqrt[3]{4 \times 10^{-7}}$

(b) $\sqrt{4 \times 10^7}$ (f) $\sqrt[3]{4 \times 10^{12}}$ (i) $\sqrt[3]{4 \times 10^{15}}$ (l) $\sqrt[3]{4 \times 10^{-8}}$

(c) $\sqrt{4 \times 10^8}$ (g) $\sqrt[3]{4 \times 10^{13}}$ (j) $\sqrt[3]{4 \times 10^{-6}}$ (m) $\sqrt[3]{4 \times 10^{-9}}$

(d) $\sqrt{4 \times 10^{-4}}$

4.37. The earth is an average of 9.3×10^7 mi from the sun. How far is this in kilometers? In meters?

4.38. How many cubic feet are there in a cubic meter?

4.39. The speed limit in many European towns is 60 km/h.   How many miles per hour is this?

4.40. The speed of light is 3.00×10^8 m/s.   What is this speed in feet per second?   In miles per second?   In miles per hour?

4.41. When the voltage across a capacitor of capacitance C is V, the charge on the capacitor is $Q = CV$.   Find the charge on a 5-μF capacitor when it is connected to a 600-V source.

4.42. A 500-pF capacitor has a charge of 0.2 μC.   What is the voltage across it?

4.43. The resonant frequency of an alternating-current (ac) circuit that contains an inductance L and a capacitance C is given by $f_0 = 1/(2\pi\sqrt{LC})$.   Find f_0 in megahertz when $L = 2$ mH   and $C = 5$ pF.

4.44. What value of C in picofarads is needed in a circuit in which $L = 50$ mH   if the resonant frequency is to be 80 kHz?

4.45. Find the logarithms of the following numbers:
(a) 2   (b) 20   (c) 200   (d) 3×10^5 (e) 6×10^8

4.46. Find the antilogarithms of the following logarithms (i.e., the numbers that correspond to these logarithms):
(a) 0.8710   (b) 1.3838   (c) 3.6484   (d) 5.5888   (e) 10

4.47. What is the power gain in decibels of an amplifier whose power input is 0.5 W and whose power output is 50 W?

4.48. What is the power gain in decibels of an amplifier whose power input is 0.15 W and whose power output is 6 W?

4.49. What is the ratio between the output and input powers of a 30-dB amplifier?

4.50. What is the ratio between the output and input powers of a 45-dB amplifier?

4.51. What is the power output when the input signal to a 40-dB amplifier is 5 mW?

4.52. A 60-dB amplifier has an output of 25 W.   What is the input power?

4.53. A 1.2-kW radio transmitter is coupled to an antenna by a cable whose attenuation is 1.8 dB.   How much power reaches the antenna?

4.54. What is the value in dBm of a 5-W signal?

4.55. What is the value in dBm of a 0.03-mW signal?

4.56. A preamplifier whose gain is 20 dB is used with a 50-dB amplifier.   What is the total gain of the system?

4.57. A microphone whose output is -40 dBm is used with the system of Problem 4.56. What is the system output in dBm? In watts?

Answers to Supplementary Problems

4.30.
(a) 7.2×10^2 (d) 6.2×10^{-5} (g) 4.9527×10^4 (j) 4.9×10^{10}
(b) 8.9×10^5 (e) 3.6×10^0 (h) 2.943×10^{-3} (k) 1.1×10^{-8}
(c) 2×10^{-2} (f) 4×10^{-1} (i) 1.4×10^{-3} (l) 1.4763×10^0

4.31.
(a) 0.0003 (d) 101 000 000 (g) 4321.45 (j) 0.000 069
(b) 7500 (e) 500 (h) 6 000 000
(c) 0.000 081 26 (f) 0.032 (i) 5.7

4.32.
(a) 4.3×10^3 (e) 2.02×10^1 (i) 2.7×10^{-4} (m) -5×10^{-5}
(b) 5.02×10^6 (f) 6.37×10^2 (j) -1.3×10^5 (n) 6.6×10^3
(c) 7.2×10^{-2} (g) 3.7×10^3 (k) -3.154×10^7 (o) -6.64×10^{-3}
(d) 5.04×10^{-3} (h) 4.991×10^7 (l) 2×10^3

4.33.
(a) 0.01 (b) 0.005 (c) 0.000 167 (d) 100 (e) 50 (f) 2500

4.34.
(a) 2.5×10^1 (d) 6×10^{15} (g) 2×10^{-8} (j) 5×10^{-6}
(b) 10^6 (e) 9×10^{-4} (h) 2.5×10^{-5}
(c) 10^3 (f) 8.3×10^{-12} (i) 6×10^{-6}

4.35.
(a) 6.4×10^{28} (c) 2.5×10^{-15} (e) 9×10^{-16} (g) 3.7×10^{10}
(b) 4×10^{14} (d) 3.2×10^{-9} (f) 4×10^{-24}

4.36.
(a) 2×10^3 (d) 2×10^{-2} (g) 3.4×10^4 (j) 1.6×10^{-2} (m) 1.6×10^{-3}
(b) 6.3×10^3 (e) 6.3×10^{-3} (h) 7.4×10^4 (k) 7.4×10^{-3}
(c) 2×10^4 (f) 1.6×10^4 (i) 1.6×10^5 (l) 3.4×10^{-3}

4.37. 1.5×10^8 km; 1.5×10^{11} m

4.38. 35.3 ft^3

4.39. 37 mi/h

4.40. 9.84×10^8 ft/s; 1.86×10^5 mi/s; 6.72×10^8 mi/h

4.41. 3×10^{-3} C = 3 mC

4.42. 400 V

4.43. 1.59 MHz

4.44. 79.2 pF

4.45. (a) 0.3010 (b) 1.3010 (c) 2.3010 (d) 5.4771 (e) 8.7782

4.46. (a) 7.43 (b) 24.2 (c) 4450 (d) 3.88×10^5 (e) 10^{10}

4.47. 20 dB   **4.50.** 31 623   **4.53.** 793 W   **4.56.** 70 dB

4.48. 16 dB   **4.51.** 50 W   **4.54.** 37 dBm   **4.57.** 30 dBm; 1 W

4.49. 1000   **4.52.** 25 μW   **4.55.** -15 dBm

Chapter 5

Resistance and Wire Size

WIRE LENGTH

The resistance R of a wire or other metallic conductor depends upon its length l, its cross-sectional area A, and the material it is made of. The resistance is directly proportional to l: if the other factors stay the same, the longer a wire is, the greater is its resistance. Doubling the length of a wire doubles its resistance and halving the length halves the resistance. Thus the ratio between the resistances R_1 and R_2 of two wires of the same kind and same cross section equals the ratio between their lengths l_1 and l_2:

$$\frac{R_1}{R_2} = \frac{l_1}{l_2}$$

To solve this equation for any one of these quantities, we keep in mind that we can shift something that multiplies one side of any equation so that it divides the other side (see Chapter 1). Similarly, we can shift a quantity that divides one side of an equation so that it multiplies the other side. Symbolically,

$$\frac{R_1}{R_2} = \frac{l_1}{l_2}$$

We can therefore solve the above equation for R_1, R_2, l_1, or l_2, with these results:

$$R_1 = \frac{l_1 R_2}{l_2} \qquad R_2 = \frac{l_2 R_1}{l_1} \qquad l_1 = \frac{R_1 l_2}{R_2} \qquad l_2 = \frac{R_2 l_1}{R_1}$$

Problem 5.1. A 20-ft length of No. 38 wire has a resistance of 13 Ω. What length of this wire will have a resistance of 8 Ω?

Let us call R_1 the resistance of the original 20-ft wire and R_2 the desired resistance of 8 Ω, so that $R_1 = 13$ Ω, $R_2 = 8$ Ω, $l_1 = 20$ ft, and $l_2 = ?$ We find from the last of the above formulas that

$$l_2 = \frac{R_2 l_1}{R_1} = \frac{8 \text{ Ω} \times 20 \text{ ft}}{13 \text{ Ω}} = 12.3 \text{ ft}$$

CROSS SECTION

The resistance of a wire varies inversely with its cross-sectional area A: if the other factors stay the same, the thicker a wire is, the less is its resistance. Doubling the area of a wire cuts its resistance in half and halving the area doubles its resistance. The ratio between the resistances R_1 and R_2 of two wires of the same kind and same length equals the inverse of the ratio of their areas A_1 and A_2:

$$\frac{R_1}{R_2} = \frac{A_2}{A_1}$$

The area of a circle whose diameter is D is

$$A = \frac{\pi D^2}{4}$$

Accordingly the above ratio of resistances can be expressed as a ratio of squared diameters when the wires in question are round:

$$\frac{R_1}{R_2} = \frac{A_2}{A_1} = \frac{\pi D_2^2/4}{\pi D_1^2/4} = \frac{D_2^2}{D_1^2}$$

Problem 5.2. A certain wire whose diameter is 0.04 in has a resistance of 2 Ω. (*a*) What would its resistance be if its diameter were 0.10 in? (*b*) If its diameter were 0.01 in?

(*a*) Here $R_1 = 2$ Ω, $D_1 = 0.04$ in, $D_2 = 0.10$ in, and $R_2 = ?$ Solving for R_2 gives

$$R_2 = \frac{D_1^2 R_1}{D_2^2} = \left(\frac{D_1}{D_2}\right)^2 R_1 = \left(\frac{0.04 \text{ in}}{0.10 \text{ in}}\right)^2 \times 2 \text{ Ω} = (0.4)^2 \times 2 \text{ Ω} = 0.16 \times 2 \text{ Ω} = 0.32 \text{ Ω}$$

(*b*) $$R_2 = \left(\frac{D_1}{D_2}\right)^2 R_1 = \left(\frac{0.04 \text{ in}}{0.01 \text{ in}}\right)^2 \times 2 \text{ Ω} = (4)^2 \times 2 \text{ Ω} = 16 \times 2 \text{ Ω} = 32 \text{ Ω}$$

Problem 5.3. What would the diameter of the above wire be if its resistance were 1 Ω?

Now $R_1 = 2$ Ω, $D_1 = 0.04$ in, $R_2 = 1$ Ω, and $D_2 = ?$ To solve for D_2, we proceed as follows:

$$D_2^2 = \frac{R_1 D_1^2}{R_2}$$

$$D_2 = \sqrt{\frac{R_1 D_1^2}{R_2}} = \sqrt{\frac{R_1}{R_2}} D_1 = \sqrt{\frac{2 \text{ Ω}}{1 \text{ Ω}}} \times 0.04 \text{ in} = \sqrt{2} \times 0.04 \text{ in} = 1.41 \times 0.04 \text{ in} = 0.056 \text{ in}$$

THE CIRCULAR MIL

In engineering practice it is customary to use as the unit of area of a round conductor the *circular mil*, or *cmil*. The *mil* is a unit of length equal to 0.001 in, which is 1/1000 in. A circular mil is a unit of area equal to the area of a circle whose diameter is 1 mil, as in Fig. 5-1. The area A_{cm} in cmils of a circle whose diameter in mils is D_m is equal to D_m^2:

$$A_{cm} = D_m^2$$

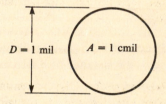

Fig. 5-1

The advantage of using the cmil as a unit of area is that it avoids multiplication and division by π.

Problem 5.4. The diameter of No. 14 wire is 0.06408 in. Find its cross-sectional area in circular mils.

Since 1 mil = 0.001 in, $D = 64.08$ mils here and

$$A_{cm} = D_m^2 = (64.08)^2 \text{ cmil} = 4106 \text{ cmil}$$

Problem 5.5. The area of No. 10 wire is 10 380 cmil. How does the resistance of a given length of No. 14 wire compare with that of the same length of No. 10 wire?

$$\frac{R_{14}}{R_{10}} = \frac{A_{10}}{A_{14}} = \frac{10\,380 \text{ cmil}}{4106 \text{ cmil}} = 2.53$$

The No. 14 wire has 2.53 times as much resistance as the No. 10 wire.

SPECIFIC RESISTANCE

Different materials differ in their ability to conduct electric current. The *specific resistance* (or *resistivity*) of a material is the resistance of a 1-ft length of it whose diameter is 1 mil, so that its cross-sectional area is 1 cmil. The usual symbol of specific resistance is ρ, the Greek letter *rho*. Because of the way ρ is defined and because the resistance R of a conductor is directly proportional to its length l and inversely proportional to its area A,

$$R = \frac{\rho l}{A}$$

The correct unit of specific resistance is the $\Omega \cdot \text{cmil/ft}$, but it is sometimes quoted as the $\Omega / \text{cmil} \cdot \text{ft}$, which is obviously wrong.

Problem 5.6. The specific resistance of the copper used in electric wires is $10.4 \, \Omega \cdot \text{cmil/ft}$. Find the resistance of 1500 ft of copper wire whose diameter is 0.080 in.

Since   0.080 in = 80 mils,

$$A_{\text{cm}} = D_{\text{m}}^2 = (80)^2 \text{ cmil} = 6400 \text{ cmil}$$

and

$$R = \frac{\rho l}{A} = \frac{10.4 \, \Omega \cdot \text{cmil/ft} \times 1500 \text{ ft}}{6400 \text{ cmil}} = 2.44 \, \Omega$$

Problem 5.7. The specific resistance of Nichrome is $600 \, \Omega \cdot \text{cmil/ft}$. How long should a Nichrome wire 20 mils in diameter be for it to have a resistance of $5 \, \Omega$?

The cross-sectional area of the wire is

$$A_{\text{cm}} = D_{\text{m}}^2 = (20)^2 \text{ cmil} = 400 \text{ cmil}$$

Next we solve for l and substitute $\rho = 600 \, \Omega \cdot \text{cmil/ft}$, $R = 5 \, \Omega$,   and $A = 400 \text{ cmil}$ to find the value of the length l:

$$l = \frac{RA}{\rho} = \frac{5 \, \Omega \times 400 \text{ cmil}}{600 \, \Omega \cdot \text{cmil/ft}} = 3.33 \text{ ft}$$

Problem 5.8. The maximum resistance of a copper wire 400 ft long is to be $1.5 \, \Omega$. What should the minimum diameter of the wire be?

The minimum area of the wire is

$$A = \frac{\rho l}{R} = \frac{10.4 \, \Omega \cdot \text{cmil/ft} \times 400 \text{ ft}}{1.5 \, \Omega} = 2773 \text{ cmil}$$

Since $A_{\text{cm}} = D_{\text{m}}^2$,

$$D_{\text{m}} = \sqrt{A_{\text{cm}}} = \sqrt{2773} \text{ mils} = 52.7 \text{ mils}$$

TEMPERATURE VARIATION

The specific resistances of most materials vary with temperature. In general, metals increase in specific resistance with an increase in temperature. If R is the resistance of a metal conductor at a certain temperature, then the change in its resistance ΔR (Δ is the Greek capital letter *delta*) when its temperature changes by ΔT is proportional to both R and ΔT, so that

$$\Delta R = \alpha R \, \Delta T$$

The quantity α (Greek letter *alpha*) is called the *temperature coefficient of resistance* and its value depends upon the metal under consideration.

Problem 5.9. A copper wire has a resistance of 10.0 Ω at 20°C. (a) What will its resistance be at 80°C? (b) At 0°C? The temperature coefficient of resistance of copper is 0.004/°C.

(a) Here $R = 10.0 \ \Omega$ and $\Delta T = 60$°C. Hence the wire's change in resistance is

$$\Delta R = \alpha R \, \Delta T = 0.004/\text{°C} \times 10.0 \ \Omega \times 60\text{°C} = 2.4 \ \Omega$$

and the resistance at 80°C will be $R + \Delta R = 12.4 \ \Omega$.

(b) Here $\Delta T = -20$°C, and so

$$\Delta R = \alpha R \, \Delta T = 0.004/\text{°C} \times 10.0 \ \Omega \times (-20\text{°C}) = -0.8 \ \Omega$$

The resistance at 0°C will be $R + \Delta R = 9.2 \ \Omega$.

Problem 5.10. A *resistance thermometer* makes use of the variation of the resistance of a conductor with temperature. If the resistance of such a thermometer with a platinum element is 5 Ω at 20°C and 16 Ω when inserted in a furnace, find the temperature of the furnace. The value of α for platinum is 0.0036/°C.

Here $R = 5 \ \Omega$ and $\Delta R = 16 \ \Omega - 5 \ \Omega = 11 \ \Omega$. Since $\Delta R = \alpha R \, \Delta T$,

$$\Delta T = \frac{\Delta R}{\alpha R} = \frac{11 \ \Omega}{0.0036/\text{°C} \times 5 \ \Omega} = 611\text{°C}$$

The temperature of the furnace is $T + \Delta T = 20\text{°C} + 611\text{°C} = 631\text{°C}$.

Problem 5.11. Motors, generators, transformers, and other electrical devices are not supposed to be operated above certain temperatures whose values depend on their construction. One way to determine the temperature in the interior of such a device is to measure the resistance of one of its windings before it is run and after it has been run for some time. If the field windings of an electric motor have a resistance of 230 Ω while the motor is operating and a resistance of 200 Ω at a room temperature of 20°C, find the operating temperature of the motor.

Here $R = 200 \ \Omega$, $\Delta R = 30 \ \Omega$, and $\alpha = 0.004/\text{°C}$ for copper wire. Solving the first equation in this section for the rise in temperature ΔT yields

$$\Delta T = \frac{\Delta R}{\alpha R} = \frac{30 \ \Omega}{0.004/\text{°C} \times 200 \ \Omega} = 37.5\text{°C}$$

so the operating temperature is $T + \Delta T = 20\text{°C} + 37.5\text{°C} = 57.5\text{°C}$. The Fahrenheit equivalent of this temperature is

$$T_\text{F} = \frac{9}{5} T_\text{C} + 32° = \frac{9}{5} (57.5°) + 32° = 103.5° + 32° = 135.5\text{°F}$$

WIRE SIZE

Wire is manufactured in the United States in the standard sizes specified by the American Wire Gage (AWG) system, which is given in Appendix B. The largest wire in this system is AWG No. 0000, which is 460 mils (nearly half an inch) in diameter, and the smallest is the hair-thin No. 40, which is 3.1 mils in diameter. The sequence of diameters is such that every third gage number means a cross-sectional area half as great and so a resistance twice as great for the same wire length. Thus No. 14 wire is 64.1 mils in diameter, 4107 cmil in area, and has a resistance of 2.53 Ω/1000 ft at 20°C according to the table, and No. 17 wire is 45.3 mils in diameter, 2048 cmil in area, and has a resistance of 5.06 Ω/1000 ft. Even sizes are normally employed for wiring purposes, with the odd sizes finding use in coils of various kinds, such as the windings of motors and transformers. Number 14 wire is the smallest permitted for residential, farm, and industrial wiring.

Problem 5.12. A bank of lamps is connected to a 230-V power source 110 ft away using No. 6 wire. If the current to the lamps is 40 A, find the voltage at the lamp terminals.

From Appendix B, No. 6 wire has a resistance of 0.395 Ω/1000 ft. The total length of wire here is

2×110 ft $= 220$ ft,   and so its resistance is

$$R = \frac{0.395 \ \Omega}{1000 \ \text{ft}} \times 220 \ \text{ft} = 0.0869 \ \Omega$$

From Ohm's law the voltage drop is

$$V = IR = 40 \ \text{A} \times 0.0869 \ \Omega = 3.5 \ \text{V}$$

The voltage at the lamp terminals is therefore $230 \ \text{V} - 3.5 \ \text{V} = 226.5 \ \text{V}$,   which represents a drop of

$$\frac{3.5 \ \text{V}}{230 \ \text{V}} = 0.015 = 1.5\%$$

AMPACITY

The *ampacity* of a wire is the maximum current it can safely carry.   When the current in a wire is larger, the heat produced in it is also larger.   If the wire gets too hot, its insulation may be damaged, so its ampacity is determined both by how its temperature varies with current and by the nature of the insulation.   The larger a wire, the greater its ampacity for a given type of insulation, but the ampacity is not proportional to the cross-sectional area, as one might think at first.   The reason is that the heat developed in a wire is dissipated through its surface, so the greater its surface, all else equal, the lower its temperature.   The surface area of a wire increases in proportion to its diameter D whereas its cross-sectional area, upon which its resistance depends, increases in proportion to D^2; the ampacity of a wire varies more closely with D than with D^2.

The National Electrical Code gives the ampacities of wires from No. 14 to No. 0000 for various kinds of insulation.   Rubber insulation of type RH can safely withstand a temperature of 75°C, for instance, and No. 14 wire with this insulation has an ampacity of 15 A in free air when the ambient temperature does not exceed 30°C.   The maximum temperature permitted for type A asbestos insulation is 200°C, and the ampacity of No. 14 asbestos-covered wire is 30 A.   An abbreviated ampacity table is given in Appendix C.

DETERMINING WIRE SIZE

Two factors govern the choice of the wire size for a particular application:

1. The maximum current in the wire
2. The maximum permissible voltage drop in the wire

The procedure is to find the smallest wire that suits each factor, and then to choose the larger of the two wires.   In the case of the first factor, one simply looks up in an ampacity table (such as that in Appendix C) the minimum wire size that can safely carry the specified current.   For the second factor, it is necessary to first calculate the highest resistance the wire can have in order not to have the voltage drop exceed the given limit.   Then the corresponding cross-sectional area of the wire is found, and this figure is used with a table of wire properties (such as that in the Appendix C) to find the minimum wire size.

Problem 5.13.   An electric motor draws 25 A from a 115-V power source 60 ft away.   What is the minimum size of rubber-insulated wire that can be used if the voltage drop is to be no more than 2 percent?

From Appendix C, No. 12 rubber-insulated wire has an ampacity of 20 A and No. 10 rubber-insulated wire has an ampacity of 30 A.   Hence the smallest wire than can be used here is No. 10 on the basis of ampacity, since No. 12 wire cannot safely carry 25 A and No. 11 wire is not normally available.

The permissible voltage drop is 2 percent of 115 V, which is

$$V = 0.02 \times 115 \ \text{V} = 2.3 \ \text{V}$$

From Ohm's law, $R = V/I$,   and so the resistance in the wire that corresponds to a voltage drop of 2.3 V when the current is 25 A is

$$R = \frac{V}{I} = \frac{2.3 \ \text{V}}{25 \ \text{A}} = 0.092 \ \Omega$$

The total length of wire needed is twice the distance between source and motor, so $l = 2 \times 60$ ft $= 120$ ft. The specific resistance of copper is $10.4\ \Omega \cdot$ cmil/ft. From $R = \rho l / A$ the area of the wire must be at least

$$A = \frac{\rho l}{R} = \frac{10.4\ \Omega \cdot \text{cmil/ft} \times 120\ \text{ft}}{0.092\ \Omega} = 13\,565\ \text{cmil}$$

According to Appendix B, No. 10 wire has an area of 10 380 cmil, so it is too small. The correct size is No. 8, whose area of 16 510 cmil is larger than the required minimum of 13 565 cmil.

Another way to arrive at the same conclusion is to note that the 120 ft of wire is to have a maximum resistance of 0.092 Ω, which corresponds to a resistance per 1000 ft of

$$\frac{0.092\ \Omega}{120\ \text{ft}}\,(1000) = \frac{0.767\ \Omega}{1000\ \text{ft}}$$

From Appendix B, No. 8 wire has a resistance of 0.628 Ω/1000 ft and is the smallest wire whose resistance does not exceed the required figure.

The various quantities used to calculate the cross-sectional area that corresponds to a given voltage drop can be combined in a single formula:

$$A\ (\text{cmil}) = \frac{10.4 \times \text{wire length (ft)} \times \text{current (A)}}{\text{voltage drop (V)}}$$

$$A_{\text{cm}} = \frac{10.4\,lI}{V}$$

Problem 5.14. What is the maximum length of No. 12 wire that can be used to carry a current of 15 A if the voltage drop is not to exceed 5 V?

We begin by looking up the area of No. 12 wire, which is 6530 cmil. Now we rewrite the preceding equation in the form

$$l = \frac{AV}{10.4\,I}$$

and substitute $A = 6530$ cmil, $V = 5$ V, and $I = 15$ A. The resulting maximum length is

$$l = \frac{AV}{10.4\,I} = \frac{6530 \times 5}{10.4 \times 15} = 209\ \text{ft}$$

SI UNITS

In the SI units discussed in Chapter 4, wire lengths are given in meters (m) and their cross-sectional areas in square millimeters (mm^2). Wire sizes are specified according to their areas, so that a particular wire might be referred to as 2.5 mm^2 instead of, say, No. 14. In these units, the specific resistance of copper at 25°C is 0.0175 $\Omega \cdot$ mm^2/m.

Problem 5.15. Find the resistance of 80 m of 2.5-mm^2 copper wire.

$$R = \frac{\rho l}{A} = \frac{0.0175\ \Omega \cdot \text{mm}^2/\text{m} \times 80\ \text{m}}{2.5\ \text{mm}^2} = 0.56\ \Omega$$

Problem 5.16. What length of 0.1-mm^2 copper wire is needed to provide a resistance of 3 Ω?

$$R = \frac{\rho l}{A}$$

$$l = \frac{RA}{\rho} = \frac{3\ \Omega \times 0.1\ \text{mm}^2}{0.0175\ \Omega \cdot \text{mm}^2/\text{m}} = 17\ \text{m}$$

Problem 5.17. The maximum allowable resistance for a 50-m length of copper wire in a certain application is 0.2 Ω. Find the minimum cross-sectional area of the wire.

$$R = \frac{\rho l}{A}$$

$$A = \frac{\rho l}{R} = \frac{0.0175 \ \Omega \cdot mm^2/m \times 50 \ m}{0.2} = 4.4 \ mm^2$$

Supplementary Problems

5.18. A 50-ft length of No. 32 copper wire has a resistance of 8.2 Ω. What length of the wire has a resistance of 5 Ω? A resistance of 20 Ω?

5.19. A 12-ft length of No. 40 copper wire has a resistance of 12.6 Ω. What is the resistance of 7 ft of this wire? Of 30 ft?

5.20. Number 24 copper wire has a diameter of 20.1 mils and a resistance of 25.67 Ω/1000 ft. Find the resistance of 200 ft of No. 30 wire, whose diameter is 10.03 mils.

5.21. Number 18 wire has a diameter of 0.0403 in. Find its diameter in mils and its cross-sectional area in circular mils.

5.22. Number 10 wire has a cross-sectional area of 10 380 cmil. Find its diameter in mils and in inches.

5.23. Find the resistance of 40 ft of copper wire whose cross-sectional area is 200 cmil. The specific resistance of copper is 10.4 $\Omega \cdot$ cmil/ft.

5.24. Find the resistance of 600 ft of copper wire whose diameter is 0.030 in.

5.25. The specific resistance of iron is 72 $\Omega \cdot$ cmil/ft. Find the resistance of 250 ft of iron wire whose diameter is 1/16 in.

5.26. The maximum resistance of a copper wire 1 mile long is to be 1 Ω. What should the minimum diameter of the wire be?

5.27. The specific resistance of platinum is 66 $\Omega \cdot$ cmil/ft. How long should a 5-mil platinum wire be in order for it to have a resistance of 30 Ω?

5.28. A copper wire has a resistance of 2.22 Ω at 25°C. What will its resistance be at 10°C? At 50°C? The temperature coefficient of resistivity of copper is 0.004/°C.

5.29. The temperature coefficient of resistivity of carbon is -0.0005/°C. If the resistance of a carbon resistor is 1000 Ω at 0°C, find its resistance at 120°C.

5.30. A resistance thermometer uses a platinum element whose resistance at 20°C is 11 Ω. When the element is placed in a furnace, its resistance increases to 33 Ω. What is the furnace temperature? The temperature coefficient of resistivity of platinum is 0.0036/°C.

5.31. Number 14 copper wire has a resistance of 2.525 Ω/1000 ft. What is the maximum length of a cable using No. 14 wire that is to connect an electric drill drawing 8 A to a 115-V outlet if the voltage at the drill is not to go below 100 V?

5.32. An electric heater drawing 25 A is connected to a 240-V outlet 80 ft away using No. 10 copper wire, whose resistance is 1.0 Ω/1000 ft. Find the voltage at the heater.

5.33. An electric pump draws 32 A from a 120-V power source 240 ft away. What is the minimum standard size of rubber-insulated wire that can be safely used? What is the voltage drop when this wire is used?

5.34. Find the resistance of 20 m of 0.8-mm² copper wire.

5.35. Find the resistance of 150 m of copper wire that is 1.0 mm in diameter.

5.36. The maximum allowable voltage drop for a certain 70-m length of cable that carries a current of 40 A is 5 V. What is the minimum cross-sectional area of the copper wires of the cable in square millimeters?

Answers to Supplementary Problems

5.18.	30.5 ft; 122 ft	**5.28.**	2.09 Ω; 2.44 Ω
5.19.	7.35 Ω; 31.5 Ω	**5.29.**	940 Ω
5.20.	20.6 Ω	**5.30.**	576°C
5.21.	40.3 mils; 1624 cmil	**5.31.**	371 ft
5.22.	101.9 mils; 0.1019 in	**5.32.**	236 V
5.23.	2.08 Ω	**5.33.**	No. 8; 9.6 V
5.24.	6.93 Ω	**5.34.**	0.44 Ω
5.25.	4.6 Ω	**5.35.**	0.84 Ω
5.26.	234 mils	**5.36.**	19.6 mm²
5.27.	11.36 ft		

Chapter 6

Series Circuits

RESISTORS IN SERIES

Two or more resistors connected one after the other, as in Fig. 6-1, are said to be in *series*. The same current I passes through each of the resistors. The *equivalent resistance* of a set of resistors connected together is the value of the single resistor that can be substituted for the entire set without changing the properties of any circuit of which the set is a part.

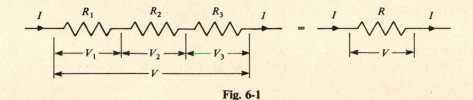

Fig. 6-1

To find the equivalent resistance of the resistors of Fig. 6-1, we start from the fact that the potential difference V across the set is the sum of the potential differences V_1, V_2, and V_3 across the individual resistors:

$$V = V_1 + V_2 + V_3$$

Because the current in each resistor is I, the potential differences across them are

$$V_1 = IR_1 \qquad V_2 = IR_2 \qquad V_3 = IR_3$$

The potential difference across the equivalent resistance R is

$$V = IR$$

Substituting for the V's in $\quad V = V_1 + V_2 + V_3 \quad$ gives

$$IR = IR_1 + IR_2 + IR_3$$

Now we divide both sides of this equation by I and find that

$$R = R_1 + R_2 + R_3$$

The same conclusion holds for any number of resistors in series: the equivalent resistance of the set is equal to the sum of the individual resistances, so we have

$$R = R_1 + R_2 + R_3 + \cdots \qquad \text{(series resistors)}$$

Problem 6.1. What is the equivalent resistance of three 5-Ω resistors connected in series?

$$R = R_1 + R_2 + R_3 = 5\ \Omega + 5\ \Omega + 5\ \Omega = 15\ \Omega$$

Problem 6.2. (*a*) If a potential difference of 60 V is applied across the above combination of resistors, as in Fig. 6-2, what is the current in each resistor? (*b*) What is the voltage drop across each resistor?

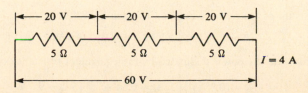

Fig. 6-2

(a) The current in the entire circuit is

$$I = \frac{V}{R} = \frac{60 \text{ V}}{15 \text{ }\Omega} = 4 \text{ A}$$

Since the resistors are in series, this current passes through each of them.

(b) The voltage drop across each of the resistors is

$$V = IR = 4 \text{ A} \times 5 \text{ }\Omega = 20 \text{ V}$$

The total voltage drop is $20 \text{ V} + 20 \text{ V} + 20 \text{ V} = 60 \text{ V}$,   which is equal to the applied potential difference.

Problem 6.3. It is desired to limit the current in a 50-Ω resistor to 10 A when it is connected to a 600-V power source.   What is the value of the series resistor that is needed?

For a current of 10 A, the equivalent resistance in the circuit should be

$$R = \frac{V}{I} = \frac{600 \text{ V}}{10 \text{ A}} = 60 \text{ }\Omega$$

Hence

$$R = R_1 + R_2$$
$$R_2 = R - R_1 = 60 \text{ }\Omega - 50 \text{ }\Omega = 10 \text{ }\Omega$$

A 10-Ω resistor should be connected in series with the 50-Ω resistor to give the required total of 60 Ω.

Problem 6.4. What is the voltage drop across each resistor in Problem 6.3?

$$V_1 = IR_1 = 10 \text{ A} \times 50 \text{ }\Omega = 500 \text{ V} \qquad V_2 = IR_2 = 10 \text{ A} \times 10 \text{ }\Omega = 100 \text{ V}$$

Problem 6.5. Two light bulbs, one of 5-Ω and the other of 10-Ω resistance, are connected in series across a potential difference of 12 V.   (a) What is the current in each bulb?   (b) What is the voltage across each bulb?   (c) What is the power dissipated by each bulb and the total power dissipated by the combination?

The solution of this problem is illustrated in Fig. 6-3.   The first step is to find the equivalent resistance of the two bulbs, which is

$$R = R_1 + R_2 = 5 \text{ }\Omega + 10 \text{ }\Omega = 15 \text{ }\Omega$$

(a) The current in the circuit is

$$I = \frac{V}{R} = \frac{12 \text{ V}}{15 \text{ }\Omega} = 0.8 \text{ A}$$

Since the bulbs are in series, the current in each of them is 0.8 A.

(b) The voltage across the 5-Ω bulb is

$$V_1 = IR_1 = 0.8 \text{ A} \times 5 \text{ }\Omega = 4 \text{ V}$$

and that across the 10-Ω bulb is

$$V_2 = IR_2 = 0.8 \text{ A} \times 10 \text{ }\Omega = 8 \text{ V}$$

As a check we note that the total voltage is

$$V = V_1 + V_2 = 4 \text{ V} + 8 \text{ V} = 12 \text{ V}$$

which equals the impressed voltage of 12 V, as it should.

(c) $\qquad P_1 = IV_1 = 0.8 \text{ A} \times 4 \text{ V} = 3.2 \text{ W} \qquad P_2 = IV_2 = 0.8 \text{ A} \times 8 \text{ V} = 6.4 \text{ W}$

The total power is the sum of the power dissipated by each bulb:

$$P = P_1 + P_2 = 3.2 \text{ W} + 6.4 \text{ W} = 9.6 \text{ W}$$

As a check, we can calculate the power dissipated by the equivalent resistance of 15 Ω:

$$P = I^2 R = (0.8 \text{ A})^2 \times 15 \text{ }\Omega = 0.64 \text{ A}^2 \times 15 \text{ }\Omega = 9.6 \text{ W}$$

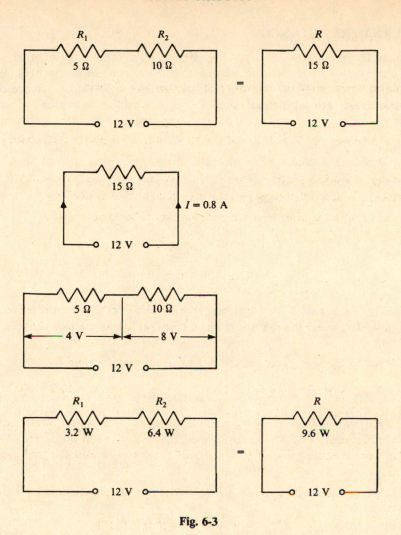

Fig. 6-3

Problem 6.6. A 2000-Ω and a 5000-Ω resistor are in series as part of a larger circuit, as in Fig. 6.4. A voltmeter shows the potential difference across the 2000-Ω resistor to be 2 V. Find the current in each resistor and the potential difference across the 5000-Ω resistor.

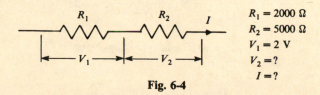

$R_1 = 2000 \ \Omega$
$R_2 = 5000 \ \Omega$
$V_1 = 2 \ \text{V}$
$V_2 = ?$
$I = ?$

Fig. 6-4

(a)   The current in the 2000-Ω resistor is

$$I = \frac{V_1}{R_1} = \frac{2 \ \text{V}}{2000 \ \Omega} = 0.001 \ \text{A}$$

This current flows through the other resistor as well.

(b)   The potential difference across R_2 is

$$V_2 = IR_2 = 0.001 \ \text{A} \times 5000 \ \Omega = 5 \ \text{V}$$

EMF AND INTERNAL RESISTANCE

The potential difference across the terminals of a battery, generator, or other source of electric energy when no current flows is called the *electromotive force* (*emf*) of the source. When a current I flows, this potential difference is less than the emf because of the *internal resistance* of the source. If the internal resistance is r, then a potential drop of Ir occurs within the source. The terminal voltage V across a source of emf V_{emf} whose internal resistance is r when it provides a current of I is therefore

$$V = V_{emf} - Ir$$

Terminal voltage = emf − potential drop due to internal resistance

When a battery or generator of emf V_{emf} is connected to an external resistance R, the total resistance in the circuit is $R + r$, and the current that flows is

$$I = \frac{V_{emf}}{R + r}$$

$$\text{Current} = \frac{\text{emf}}{\text{external resistance} + \text{internal resistance}}$$

Problem 6.7. A dry cell of emf 1.5 V and internal resistance 0.05 Ω is connected to a flashlight bulb whose resistance is 0.4 Ω. Find the current in the circuit.

$$I = \frac{V_{emf}}{R + r} = \frac{1.5 \text{ V}}{0.4 \text{ Ω} + 0.05 \text{ Ω}} = 3.33 \text{ A}$$

Problem 6.8. A battery whose emf is 45 V is connected to a 20-Ω resistance and a current of 2.1 A flows. (*a*) Find the internal resistance of the battery. (*b*) Find the terminal voltage of the battery.

(*a*) From $I = V_{emf}/(R + r)$ we obtain

$$r = \frac{V_{emf}}{I} - R = \frac{45 \text{ V}}{2.1 \text{ A}} - 20 \text{ Ω} = 21.4 \text{ Ω} - 20 \text{ Ω} = 1.4 \text{ Ω}$$

(*b*) $$V = V_{emf} - Ir = 45 \text{ V} - (2.1 \text{ A} \times 1.4) = 42 \text{ V}$$

Problem 6.9. A generator has an emf of 120 V and an internal resistance of 0.2 Ω. (*a*) How much current does the generator supply when the terminal voltage is 115 V? (*b*) How much power does it supply? (*c*) How much power is dissipated in the generator itself?

(*a*) From $V = V_{emf} - Ir$ we obtain

$$I = \frac{V_{emf} - V}{r} = \frac{120 \text{ V} - 115 \text{ V}}{0.2 \text{ Ω}} = 25 \text{ A}$$

(*b*) $$P = IV = 25 \text{ A} \times 115 \text{ V} = 2875 \text{ W}$$

(*c*) $$P = I^2 r = (25 \text{ A})^2 \times 0.2 \text{ Ω} = 125 \text{ W}$$

Problem 6.10. A source of what potential difference is required in order to charge a battery of $V_{emf} = 6$ V and $r = 0.1$ Ω at a rate of 10 A?

The required potential difference must equal the emf of the battery *plus* the Ir drop in its internal resistance. Hence

$$V_{applied} = V_{emf} + Ir = 6 \text{ V} + (10 \text{ A} \times 0.1 \text{ Ω}) = 7 \text{ V}$$

Problem 6.11. When a source of emf whose internal resistance is r is connected to an external load of resistance R, the power delivered to R will be a maximum when $R = r$. Verify this statement by calculating the power delivered by a battery of emf 10 V and internal resistance 0.5 Ω when it is connected to (a) 0.25-Ω, (b) 0.5-Ω, and (c) 1-Ω resistors.

(a) $\quad I_1 = \dfrac{V_{emf}}{R_1 + r} = \dfrac{10\text{ V}}{0.25\ \Omega + 0.5\ \Omega} = 13.3\text{ A}$ $\qquad P_1 = I_1^2 R_1 = (13.3\text{ A})^2 \times 0.25\ \Omega = 44\text{ W}$

(b) $\quad I_2 = \dfrac{V_{emf}}{R_2 + r} = \dfrac{10\text{ V}}{0.5\ \Omega + 0.5\ \Omega} = 10\text{ A}$ $\qquad P_2 = I_2^2 R_2 = (10\text{ A})^2 \times 0.5\ \Omega = 50\text{ W}$

(c) $\quad I_3 = \dfrac{V_{emf}}{R_3 + r} = \dfrac{10\text{ V}}{1\ \Omega + 0.5\ \Omega} = 6.67\text{ A}$ $\qquad P_3 = I_3^2 R_3 = (6.67\text{ A})^2 \times 1\ \Omega = 44\text{ W}$

VOLTAGE DIVIDERS

An electronic device such as an amplifier or a radio or television receiver is usually supplied with direct current that has been rectified and filtered from an alternating current source. The output of the power supply is normally at a voltage that corresponds to the highest one required by the device, and a *voltage divider* is used to obtain the other voltages needed.

The simple voltage dividers shown in Fig. 6-5 consist of several resistors connected in series (or a single resistor tapped at various points) across the outputs of their respective power supplies. The voltage divider of Fig. 6-5(a) has three equal resistors and the no-load voltage across each resistor is just 1/3 of the total voltage of the power supply. If the resistors are different, as in Fig. 6-5(b), the voltage steps are different, but of course add up to the voltage of the power supply. The voltage divider of Fig. 6-5(c) provides both positive and negative voltages by virtue of being grounded at an intermediate point rather than at one end of the series of resistors. When these voltage dividers are connected to loads, the voltages at their intermediate terminals will differ from those shown by amounts that depend upon the currents drawn.

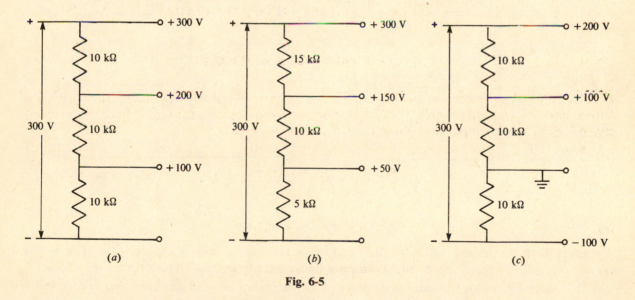

Fig. 6-5

Problem 6.12. A voltage divider is to provide currents of 20 mA (1 mA = 0.001 A, so 20 mA = 0.02 A) at 200 V, 40 mA at 100 V, and 10 mA at 60 V, as in Fig. 6-6. Find the values of the three resistors required. Assume that the current through R_1 (the *bleeder current*) is 10 percent of the total of the load currents, which is a common proportion.

(a) The total of the load currents is   20 mA + 40 mA + 10 mA = 70 mA. The bleeder current I_1 is 10 percent of this, so

$$I_1 = 0.1 \times 70\text{ mA} = 7\text{ mA} = 0.007\text{ A}$$

The potential difference between A and B is 60 V, hence

$$R_1 = \frac{V_{AB}}{I_1} = \frac{60\text{ V}}{0.007\text{ A}} = 8571\ \Omega$$

(b) The current I_2 is the sum of the 10-mA current to the load at B and the bleeder current I_1:

$$I_2 = 10\text{ mA} + I_1 = 10\text{ mA} + 7\text{ mA} = 17\text{ mA} = 0.017\text{ A}$$

The potential difference between B and C is

$$V_{BC} = V_C - V_B = 100\text{ V} - 60\text{ V} = 40\text{ V}$$

and so

$$R_2 = \frac{V_{BC}}{I_2} = \frac{40\text{ V}}{0.017\text{ A}} = 2353\ \Omega$$

(c) The current I_3 is the sum of the 40-mA current to the load at C and the current I_2:

$$I_3 = 40\text{ mA} + I_2 = 40\text{ mA} + 17\text{ mA} = 57\text{ mA} = 0.057\text{ A}$$

The potential difference between C and D is

$$V_{CD} = V_D - V_C = 200\text{ V} - 100\text{ V} = 100\text{ V}$$

and so

$$R_3 = \frac{V_{CD}}{I_3} = \frac{100\text{ V}}{0.057\text{ A}} = 1754\ \Omega$$

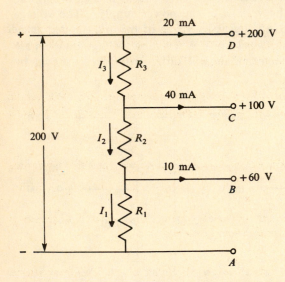

Fig. 6-6

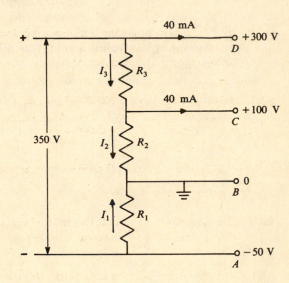

Fig. 6-7

Problem 6.13. Find the values of the resistors in the voltage divider of Fig. 6-7. The -50-V terminal provides a bias voltage and no current is drawn from it, and the bleeder current I_2 is 8 mA.

(a) The current I_1 equals the total current provided by the power supply, which is the total load current of $(40 + 40)$ mA = 80 mA   plus the bleeder current of 8 mA.   Thus

$$I_1 = 80\text{ mA} + 8\text{ mA} = 88\text{ mA} = 0.088\text{ A}$$

and, since the potential difference between A and B is 50 V,

$$R_1 = \frac{V_{AB}}{I_1} = \frac{50\text{ V}}{0.088\text{ A}} = 568\ \Omega$$

(b) The current I_2 is the bleeder current of 8 mA, so

$$I_2 = 8\text{ mA} = 0.008\text{ A}$$

Since the potential difference between B and C is

$$V_{BC} = V_C - V_B = 100 \text{ V} - 0 = 100 \text{ V}$$

the required resistance R_2 is

$$R_2 = \frac{V_{BC}}{I_2} = \frac{100 \text{ V}}{0.008 \text{ A}} = 12\,500 \text{ } \Omega$$

(c) The current I_3 is the sum of the 40-mA current to the load at C plus the bleeder current I_2 of 8 mA:

$$I_3 = 40 \text{ mA} + 8 \text{ mA} = 48 \text{ mA} = 0.048 \text{ A}$$

Since the potential difference between C and D is

$$V_{CD} = V_D - V_C = 300 \text{ V} - 100 \text{ V} = 200 \text{ V}$$

the required resistance R_3 is

$$R_3 = \frac{V_{CD}}{I_3} = \frac{200 \text{ V}}{0.048 \text{ A}} = 4167 \text{ } \Omega$$

Supplementary Problems

6.14. (a) Find the equivalent resistance of four 60-Ω resistors connected in series. (b) If a potential difference of 12 V is applied across the combination, what is the current in each resistor? (c) What is the voltage drop across each resistor? (d) How much power does each resistor dissipate?

6.15. A 100-Ω resistor and a 200-Ω resistor are connected in series with a 40-V power source. (a) What is the current in each resistor? (b) What is the potential difference across each resistor? (c) How much power does each resistor dissipate?

6.16. A 5-Ω resistor and a 20-Ω resistor are connected in series with a 100-V power source. Find the power each resistor dissipates.

6.17. A 25-Ω, a 40-Ω, and a 60-Ω resistor are in series in a circuit such that the voltage across the 25-Ω resistor is 18 V. Find the voltage across the other resistors and the current in each of them.

6.18. It is desired to limit the current in a 10-Ω resistor to 5 A when it is connected to a 120-V power source. What series resistor is required?

6.19. A dry cell has an emf of 1.5 V and an internal resistance of 0.08 Ω. (a) Find the current when the cell's terminals are connected together. (b) Find the current when the cell is connected to a 5-Ω resistance.

6.20. A certain "12-V" storage battery actually has an emf of 13.2 V and an internal resistance of 0.01 Ω. What is the terminal voltage of the battery when it delivers 80 A to the starter motor of a car engine?

6.21. A generator whose emf is 240 V has a terminal voltage of 220 V when it delivers a current of 50 A. (a) Find the internal resistance of the generator. (b) Find the power supplied by the generator. (c) Find the power dissipated within the generator.

6.22. A storage battery of emf 34 V and internal resistance 0.1 Ω is to be charged at a rate of 20 A from a 110-V source. What series resistance is needed in the circuit?

6.23. A current of 2.2 A flows when a battery of emf 24 V is connected to a 10-Ω load.   Find (a) the internal resistance of the battery and (b) its terminal voltage.

6.24. Design a voltage divider that will furnish 10 mA at 300 V, 20 mA at 250 V, and 30 mA at 50 V.   Provide a bleeder current of 5 mA.

6.25. Design a voltage divider that will furnish 50 mA at 200 V, 100 mA at 100 V, and 50 mA at 50 V.   Provide a bleeder current of 20 mA.

6.26. Design a voltage divider that will furnish 50 mA at 150 V, at 100 V, and at 50 V.   Provide a bleeder current of 15 mA.

Answers to Supplementary Problems

6.14. (a)   240 Ω (b)   0.05 A   (c)   3 V   (d)   0.15 W

6.15. (a)   0.133 A; 0.133 A   (b)   13.3 V; 26.7 V   (c)   1.77 W; 3.54 W

6.16. 80 W; 320 W

6.17. (a)   28.8 V across the 40-Ω resistor; 43.2 V across the 60-Ω resistor   (b)   0.72 A in each resistor

6.18. 14 Ω

6.19. (a)   19 A   (b)   0.3 A

6.20. 12.4 V

6.21. (a)   0.4 Ω (b)   11 kW   (c)   1 kW

6.22. 3.7 Ω

6.23. (a)   0.9 Ω (b)   22 V

6.24. In sequence from the $+300$-V terminal, the required resistances are $R_3 = 909\ \Omega$, $R_2 = 5714\ \Omega$,   and $R_1 = 10\,000\ \Omega$.

6.25. In sequence from the $+200$-V terminal, the required resistances are $R_3 = 588\ \Omega$, $R_2 = 714\ \Omega$,   and $R_1 = 2500\ \Omega$.

6.26. In sequence from the $+150$-V terminal, the required resistances are 435 Ω, 769 Ω, and 3333 Ω.

Chapter 7

Parallel Circuits

RESISTORS IN PARALLEL

In a *parallel* set of resistors, the corresponding terminals of the resistors are connected together (Fig. 7-1). The same potential difference V is present across all of them and the total current I is divided among them. As in the case of resistors in series, the equivalent resistance of the set is the value of the single resistor that can be substituted for the entire set without changing the properties of any circuit of which the set is a part.

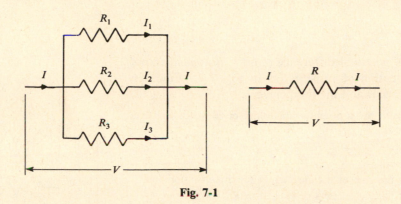

Fig. 7-1

To find the equivalent resistance of the resistors of Fig. 7-1, we start from the fact that the total current I is equal to the sum of the currents through the separate resistors:

$$I = I_1 + I_2 + I_3$$

Because the potential difference V is the same across all the resistors, their respective currents are

$$I_1 = \frac{V}{R_1} \qquad I_2 = \frac{V}{R_2} \qquad I_3 = \frac{V}{R_3}$$

The smaller the resistance, the greater the current through a resistor in a parallel set. The total current is given in terms of the equivalent resistance R by

$$I = \frac{V}{R}$$

Substituting for the I's in $I = I_1 + I_2 + I_3$ gives

$$\frac{V}{R} = \frac{V}{R_1} + \frac{V}{R_2} + \frac{V}{R_3}$$

Now we divide both sides of this equation by V, bearing in mind that $V/V = 1$. The result is

$$\frac{1}{R} = \frac{1}{R_1} + \frac{1}{R_2} + \frac{1}{R_3}$$

The same conclusion holds for any number of resistors in parallel: the reciprocal $1/R$ of the equivalent resistance of the set is equal to the sum of the reciprocals of the individual resistances, so we have

$$\frac{1}{R} = \frac{1}{R_1} + \frac{1}{R_2} + \frac{1}{R_3} + \cdots \quad \text{(Parallel resistors)}$$

59

Problem 7.1. What is the equivalent resistance of three 5-Ω resistors connected in parallel?

$$\frac{1}{R} = \frac{1}{R_1} + \frac{1}{R_2} + \frac{1}{R_3} = \frac{1}{5\ \Omega} + \frac{1}{5\ \Omega} + \frac{1}{5\ \Omega}$$

When fractions to be added have the same denominator, their sum is the sum of their numerators (upper parts) with the same denominator (lower part), as discussed in Chap. 2:

$$\frac{a}{d} + \frac{b}{d} + \frac{c}{d} = \frac{a+b+c}{d}$$

Here the fractions on the right-hand side of the equation all have the denominator 5 Ω, so

$$\frac{1}{R} = \frac{1}{5\ \Omega} + \frac{1}{5\ \Omega} + \frac{1}{5\ \Omega} = \frac{1+1+1}{5\ \Omega} = \frac{3}{5\ \Omega}$$

Finally we cross-multiply to obtain the value of R:

$$\frac{1}{R} = \frac{3}{5\ \Omega}$$

$$5\ \Omega = 3R$$

$$R = \frac{5\ \Omega}{3} = \frac{5}{3}\ \Omega = 1.67\ \Omega$$

Problem 7.2. A potential difference of 60 V is applied across the above combination of resistors, as in Fig. 7-2. Find the current in each resistor and the total current.

Since each resistor has a potential difference of 60 V across it, the current in each one is

$$I = \frac{V}{R} = \frac{60\ V}{5\ \Omega} = 12\ A$$

The total current is the sum of the currents in the three resistors:

$$I = I_1 + I_2 + I_3 = 12\ A + 12\ A + 12\ A = 36\ A$$

As a check, we can find the total current by using the equivalent resistance of 1.67 Ω:

$$I = \frac{V}{R} = \frac{60\ V}{1.67\ \Omega} = 36\ A$$

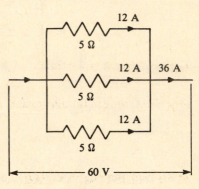

Fig. 7-2

Problem 7.3. Find the equivalent resistance of a 20-Ω resistor, a 40-Ω resistor, and a 50-Ω resistor connected in parallel.

$$\frac{1}{R} = \frac{1}{R_1} + \frac{1}{R_2} + \frac{1}{R_3} = \frac{1}{20\ \Omega} + \frac{1}{40\ \Omega} + \frac{1}{50\ \Omega}$$

The fractions on the right have different denominators. In order to add them, they must have a common denominator. We note that the denominators are all factors of 200, so this is a suitable common denominator. The fractions are converted to a denominator of 200 as follows:

$$\frac{1}{20\ \Omega} \times \frac{10}{10} = \frac{10}{200\ \Omega} \qquad \frac{1}{40\ \Omega} \times \frac{5}{5} = \frac{5}{200\ \Omega} \qquad \frac{1}{50\ \Omega} \times \frac{4}{4} = \frac{4}{200\ \Omega}$$

Hence

$$\frac{1}{R} = \frac{10}{200\ \Omega} + \frac{5}{200\ \Omega} + \frac{4}{200\ \Omega} = \frac{10+5+4}{200\ \Omega} = \frac{19}{200\ \Omega}$$

As in Problem 7.1 we cross-multiply to find R:

$$\frac{1}{R} = \frac{19}{200\ \Omega}$$

$$200\ \Omega = 19R$$

$$R = \frac{200\ \Omega}{19} = \frac{200}{19}\ \Omega = 10.5\ \Omega$$

Another approach, which is often easier, is to convert each fraction to a decimal and proceed from there. We have

$$\frac{1}{20\ \Omega} = \frac{1}{20} \times \frac{1}{\Omega} = 0.05\ \frac{1}{\Omega} \qquad \frac{1}{40\ \Omega} = \frac{1}{40} \times \frac{1}{\Omega} = 0.025\ \frac{1}{\Omega} \qquad \frac{1}{50\ \Omega} = \frac{1}{50} \times \frac{1}{\Omega} = 0.02\ \frac{1}{\Omega}$$

Hence

$$\frac{1}{R} = (0.05 + 0.025 + 0.02)\,\frac{1}{\Omega} = 0.095\,\frac{1}{\Omega}$$

and so, cross-multiplying,

$$1\,\Omega = 0.095\,R$$

$$R = \frac{1}{0.095}\,\Omega = 10.5\,\Omega$$

Problem 7.4. A potential difference of 60 V is applied across the above combination of resistors, as in Fig. 7-3. Find the current in each resistor and the total current.

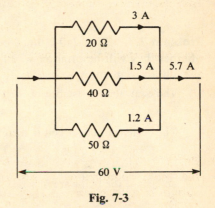

The currents in the various resistors are as follows:

$$I_1 = \frac{V}{R_1} = \frac{60\text{ V}}{20\ \Omega} = 3\text{ A}$$

$$I_2 = \frac{V}{R_2} = \frac{60\text{ V}}{40\ \Omega} = 1.5\text{ A}$$

$$I_3 = \frac{V}{R_3} = \frac{60\text{ V}}{50\ \Omega} = 1.2\text{ A}$$

Fig. 7-3

The total current is therefore

$$I = I_1 + I_2 + I_3 = 3\text{ A} + 1.5\text{ A} + 1.2\text{ A} = 5.7\text{ A}$$

As a check, we can find the total current by using the equivalent resistance of 10.5 Ω:

$$I = \frac{V}{R} = \frac{60\text{ V}}{10.5\ \Omega} = 5.7\text{ A}$$

TWO RESISTORS IN PARALLEL

This equivalent resistance of two resistors connected in parallel is given by

$$\frac{1}{R} = \frac{1}{R_1} + \frac{1}{R_2}$$

The lowest common denominator of the right-hand side is $R_1 R_2$, and so we can express the fractions $1/R_1$ and $1/R_2$ as

$$\frac{1}{R_1} \times \frac{R_2}{R_2} = \frac{R_2}{R_1 R_2} \qquad \frac{1}{R_2} \times \frac{R_1}{R_1} = \frac{R_1}{R_1 R_2}$$

Hence we have

$$\frac{1}{R} = \frac{R_2}{R_1 R_2} + \frac{R_1}{R_1 R_2} = \frac{R_1 + R_2}{R_1 R_2}$$

Inverting both sides of this equation gives the convenient formula

$$R = \frac{R_1 R_2}{R_1 + R_2}$$

Problem 7.5. Find the equivalent resistance of a 5-Ω resistor and a 10-Ω resistor connected in parallel.

$$R = \frac{R_1 R_2}{R_1 + R_2} = \frac{5\ \Omega \times 10\ \Omega}{5\ \Omega + 10\ \Omega} = \frac{50\ \Omega^2}{15\ \Omega} = 3.33\ \Omega$$

Problem 7.6. A circuit has a resistance of 50 Ω. How can it be reduced to 20 Ω?

To obtain an equivalent resistance of $R = 20\ \Omega$, a resistor R_2 must be connected in parallel with the circuit of $R_1 = 50\ \Omega$. To find R_2 we proceed as follows:

$$\frac{1}{R} = \frac{1}{R_1} + \frac{1}{R_2}$$

$$\frac{1}{R_2} = \frac{1}{R} - \frac{1}{R_1} = \frac{R_1 - R}{R_1 R}$$

$$R_2 = \frac{R_1 R}{R_1 - R} = \frac{50\ \Omega \times 20\ \Omega}{50\ \Omega - 20\ \Omega} = 33.3\ \Omega$$

Problem 7.7. Two 240-Ω light bulbs are to be connected to a 120-V power source. To determine whether they will be brighter when connected in series or in parallel, calculate the power they dissipate in each arrangment.

(*a*) The equivalent resistance of the two bulbs when they are in series is

$$R = R_1 + R_2 = 240\ \Omega + 240\ \Omega = 480\ \Omega$$

The current in the circuit is therefore

$$I = \frac{V}{R} = \frac{120\ V}{480\ \Omega} = 0.25\ A$$

and, since the bulbs are in series, this current passes through each of them. The power each bulb dissipates is

$$P = I^2 R = (0.25\ A)^2 \times 240\ \Omega = 15\ W$$

(*b*) When the bulbs are in parallel, the potential difference across each of them is 120 V. Hence the power each bulb dissipates is

$$P = \frac{V^2}{R} = \frac{(120\ V)^2}{240\ \Omega} = 60\ W$$

The bulbs will be brighter when connected in parallel.

SERIES-PARALLEL COMBINATIONS

Sometimes a set of resistors in parallel is connected in series with one or more other resistors. The procedure for finding the equivalent resistance of such a combination is to first find the equivalent resistance of the parallel set and then add this to the resistance of the series resistors. In a complicated network of resistors this operation may have to be repeated several times.

Problem 7.8. Find the equivalent resistance of the circuit shown in Fig. 7-4.

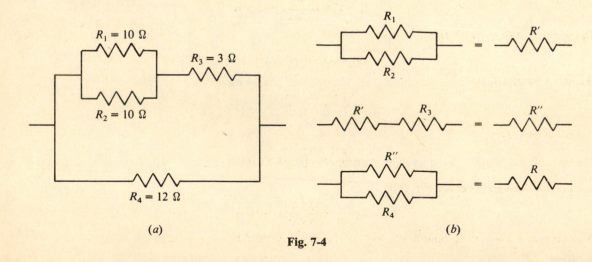

(*a*)   (*b*)

Fig. 7-4

Figure 7-4(b) shows how the original circuit is decomposed into its series and parallel parts, each of which is treated in turn.   The equivalent resistance of R_1 and R_2 is

$$R' = \frac{R_1 R_2}{R_1 + R_2} = \frac{10\ \Omega \times 10\ \Omega}{10\ \Omega + 10\ \Omega} = 5\ \Omega$$

This equivalent resistance is in series with R_3, and so

$$R'' = R' + R_3 = 5\ \Omega + 3\ \Omega = 8\ \Omega$$

Finally R'' is in parallel with R_4, hence the equivalent resistance of the entire circuit is

$$R = \frac{R'' R_4}{R'' + R_4} = \frac{8\ \Omega \times 12\ \Omega}{8\ \Omega + 12\ \Omega} = 4.8\ \Omega$$

Problem 7.9.   A potential difference of 20 V is applied to the circuit of Fig. 7-5.   Find the current through each resistor and the current through the entire circuit.

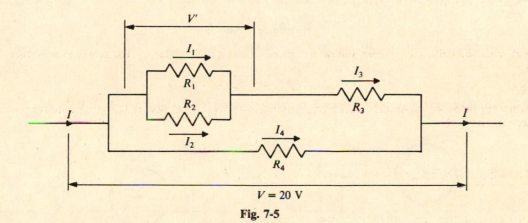

$$V = 20\ V$$

Fig. 7-5

Because resistor R_4 has the full 20-V potential difference across it,

$$I_4 = \frac{V}{R_4} = \frac{20\ V}{12\ \Omega} = 1.67\ A$$

From Fig. 7-5 we see that the current I_3 flows through the entire upper branch of the circuit, whose equivalent resistance is $R'' = 8\ \Omega$.   Hence

$$I_3 = \frac{V}{R''} = \frac{20\ V}{8\ \Omega} = 2.5\ A$$

The potential difference V' across R_1 and R_2 is

$$V' = V - I_3 R_3 = 20\ V - (2.5\ A \times 3\ \Omega) = 12.5\ V$$

Hence the current I_1 is

$$I_1 = \frac{V'}{R_1} = \frac{12.5\ V}{10\ \Omega} = 1.25\ A$$

and the current I_2 is

$$I_2 = \frac{V'}{R_2} = \frac{12.5\ V}{10\ \Omega} = 1.25\ A$$

The current through the entire circuit is

$$I = \frac{V}{R} = \frac{20\ V}{4.8\ \Omega} = 4.17\ A$$

We note that $I = I_3 + I_4$ and that $I_3 = I_1 + I_2$,   as they should.

AMMETERS AND VOLTMETERS

An *ammeter* is an instrument that measures current. The lower the resistance of an ammeter, the better, since this resistance affects the circuit whose current is being measured. In practice, the meter itself (usually a *galvanometer* in which magnetic forces produced by a current rotate a pointer) is used in parallel with a low-resistance *shunt* which carries nearly all the current, leaving a small fraction to pass through the higher-resistance meter.

A *voltmeter* is an instrument that measures potential difference. The higher the resistance of a voltmeter, the better, since its presence across a circuit element reduces the current through that element and thus changes the potential difference being measured. In practice, the meter itself (again it is usually a galvanometer) is used in series with a high resistance.

Problem 7.10. A galvanometer which measures currents from 0 to 1 mA (1 mA = 1 milliampere = 0.001 A) has a resistance of 40 Ω. How can this galvanometer be used to measure currents from 0 to 1 A?

What is needed here is a shunt resistor that will carry 0.999 A when the total current is 1.000 A (Fig. 7-6). To find the value of the shunt resistance, we note that the potential difference V across both R_{meter} and R_{shunt} is the same, so that

$$V = I_{meter} R_{meter} = I_{shunt} R_{shunt}$$

Since the meter current is to be 0.001 A when $I_{shunt} = 0.999$ A,

$$R_{shunt} = \frac{I_{meter}}{I_{shunt}} \times R_{meter} = \frac{0.001 \text{ A}}{0.999 \text{ A}} \times 40 \text{ } \Omega = 0.04 \text{ } \Omega$$

A 0.04-Ω resistor in parallel with the meter will permit it to measure currents from 0 to 1 A.

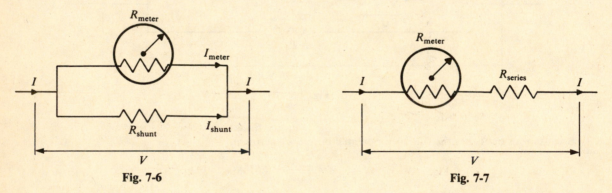

Fig. 7-6   Fig. 7-7

Problem 7.11. The galvanometer of Problem 7.10 is to be used to measure potential differences from 0 to 1 V. How can this be done?

What is needed now is a resistor in series with the meter that will limit the current to 0.001 A when the applied potential difference is 1 V (Fig. 7-7). The equivalent resistance of meter and resistor must therefore be

$$R = R_{meter} + R_{series} = \frac{V}{I}$$

and so

$$R_{series} = \frac{V}{I} - R_{meter} = \frac{1 \text{ V}}{0.001 \text{ A}} - 40 \text{ } \Omega = 960 \text{ } \Omega$$

A 960-Ω resistor in series with the meter will permit it to measure potential differences from 0 to 1 V.

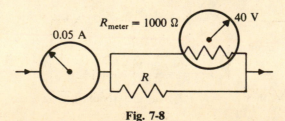

Fig. 7-8

Problem 7.12. A voltmeter whose resistance is 1000 Ω is connected across a resistor and the combination is connected in series with an ammeter (Fig. 7-8). When a potential difference is applied, the voltmeter reads 40 V and the ammeter reads 0.05 A. What is the resistance of the resistor?

At first glance it would seem that the resistance is simply

$$R' = \frac{V}{I} = \frac{40 \text{ V}}{0.05 \text{ A}} = 800 \text{ } \Omega$$

However, the voltmeter's own resistance is significant here since some of the current in the circuit is diverted through it, and $R' = 800$ Ω is actually the equivalent resistance of R and R_{meter} in parallel. Hence

$$\frac{1}{R'} = \frac{1}{R} + \frac{1}{R_{\text{meter}}}$$

$$\frac{1}{R} = \frac{1}{R'} - \frac{1}{R_{\text{meter}}} = \frac{R_{\text{meter}} - R'}{R_{\text{meter}} R'}$$

$$R = \frac{R_{\text{meter}} R'}{R_{\text{meter}} - R'} = \frac{1000 \text{ } \Omega \times 800 \text{ } \Omega}{1000 \text{ } \Omega - 800 \text{ } \Omega} = 4000 \text{ } \Omega$$

STAR AND DELTA CIRCUITS

The three resistors R_A, R_B, and R_C of Fig. 7-9 are said to form a *star* or *Y* circuit. The resistance between any two of the terminals of the circuit is simply the sum of the resistances between them. Thus the resistance between A and B is

$$R_{AB} = R_A + R_B$$

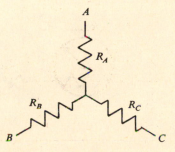

Fig. 7-9

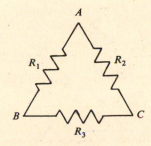

Fig. 7-10

Three resistors connected to make up a triangle, as in Fig. 7-10, are said to form a *delta* circuit because the Greek capital letter delta (Δ) has this shape. The resistance between any two of the circuit terminals is more complicated to figure out because all three resistors are involved. To find the resistance R_{AB} between A and B, the circuit of Fig. 7-10 is first redrawn as in Fig. 7-11, from which it is clear that R_1 is in parallel with the series combination of R_2 and R_3. Thus

$$R_{AB} = \frac{R_1 \times (R_2 + R_3)}{R_1 + (R_2 + R_3)} = \frac{R_1(R_2 + R_3)}{R_1 + R_2 + R_3}$$

Every delta circuit has a star circuit equivalent to it in the sense that the resistance between any pair of terminals is the same in both. Replacing a delta circuit by its star equivalent often simplifies

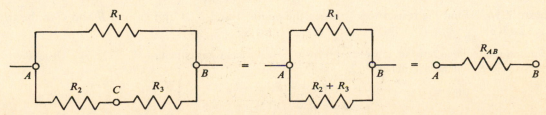

Fig. 7-11

calculations considerably.   (The reverse transformation, from star to delta, is also always possible, but offers no advantage.)   Routine algebra shows that the values of the resistors in the star circuit of Fig. 7-9 that make it equivalent to the delta circuit of Fig. 7-10 are

$$R_A = \frac{R_1 R_2}{R_1 + R_2 + R_3} \qquad R_B = \frac{R_1 R_3}{R_1 + R_2 + R_3} \qquad R_C = \frac{R_2 R_3}{R_1 + R_2 + R_3}$$

If we draw a star circuit inside the delta circuit it is to replace, as in Fig. 7-12, we see that the numerator in the formula for each star resistance is the product of the delta resistances on both sides of it.   This observation makes it easy to arrive at the correct formula in a given case.

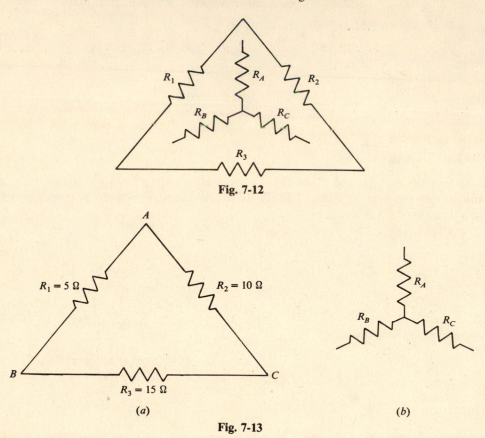

Fig. 7-12

Fig. 7-13

Problem 7-13.   Find the values of the resistors in the star circuit that is equivalent to the delta circuit of Fig. 7-13(a).

$$R_A = \frac{R_1 R_2}{R_1 + R_2 + R_3} = \frac{5\ \Omega \times 10\ \Omega}{5\ \Omega + 10\ \Omega + 15\ \Omega} = \frac{50}{30}\ \Omega = 1.67\ \Omega$$

$$R_B = \frac{R_1 R_3}{R_1 + R_2 + R_3} = \frac{5\ \Omega \times 15\ \Omega}{5\ \Omega + 10\ \Omega + 15\ \Omega} = \frac{75}{30}\ \Omega = 2.5\ \Omega$$

$$R_C = \frac{R_2 R_3}{R_1 + R_2 + R_3} = \frac{10\ \Omega \times 15\ \Omega}{5\ \Omega + 10\ \Omega + 15\ \Omega} = \frac{150}{30}\ \Omega = 5\ \Omega$$

Problem 7.14.   Find the resistances between the points A and B, A and C, and B and C in the circuit of Fig. 7-13(a).

To find the required resistances we must refer to the equivalent star circuit of Fig. 7-13(b).   Using the values of R_A, R_B, and R_C found in Problem 7.13,

$$R_{AB} = R_A + R_B = 1.67\ \Omega + 2.5\ \Omega = 4.17\ \Omega$$

$$R_{AC} = R_A + R_C = 1.67\ \Omega + 5\ \Omega = 6.67\ \Omega$$

$$R_{BC} = R_B + R_C = 2.5\ \Omega + 5\ \Omega = 7.5\ \Omega$$

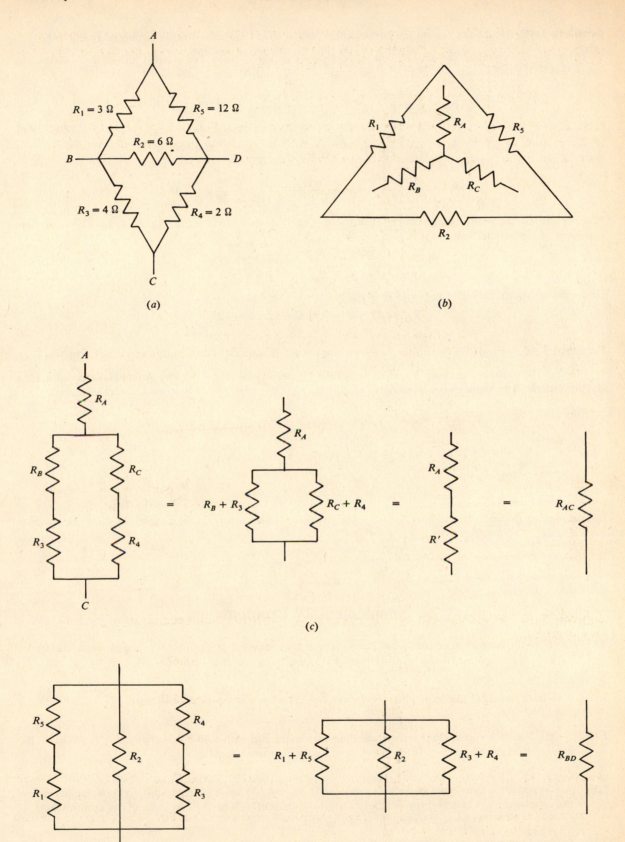

$R_1 = 3\ \Omega$ $R_5 = 12\ \Omega$

$R_2 = 6\ \Omega$

$R_3 = 4\ \Omega$ $R_4 = 2\ \Omega$

(a)

(b)

(c)

(d)

Fig. 7-14

Problem 7.15. Find the resistance between the points A and C in the bridge circuit of Fig. 7-14(a).

The star equivalent of the delta circuit ABD is shown in Fig. 7-14(b). (We could just as well use the star equivalent of the delta circuit BCD.) The resistances R_A, R_B, and R_C are as follows:

$$R_A = \frac{R_1 R_5}{R_1 + R_2 + R_5} = \frac{3\ \Omega \times 12\ \Omega}{3\ \Omega + 6\ \Omega + 12\ \Omega} = \frac{36}{21}\ \Omega = 1.71\ \Omega$$

$$R_B = \frac{R_1 R_2}{R_1 + R_2 + R_5} = \frac{3\ \Omega \times 6\ \Omega}{3\ \Omega + 6\ \Omega + 12\ \Omega} = \frac{18}{21}\ \Omega = 0.86\ \Omega$$

$$R_C = \frac{R_2 R_5}{R_1 + R_2 + R_5} = \frac{6\ \Omega \times 12\ \Omega}{3\ \Omega + 6\ \Omega + 12\ \Omega} = \frac{72}{21}\ \Omega = 3.43\ \Omega$$

Now we redraw the entire bridge circuit using the star equivalent of ABD, as in Fig. 7-14(c). The equivalent resistance of the parallel resistors is

$$R' = \frac{(R_B + R_3) \times (R_C + R_4)}{(R_B + R_3) + (R_C + R_4)} = \frac{4.86\ \Omega \times 5.43\ \Omega}{4.86\ \Omega + 5.43\ \Omega} = 2.56\ \Omega$$

Hence the resistance between the points A and C is

$$R_{AC} = R_A + R' = 1.71\ \Omega + 2.56\ \Omega = 4.27\ \Omega$$

Problem 7.16. Find the resistance between the points B and D of the bridge circuit of Fig. 7-14(a).

To find the resistance between B and D, we redraw the circuit as in Fig. 7-14(d). Although we could use the star equivalent of the delta circuit, there is no need to do so. We proceed as for any other parallel circuit:

$$\frac{1}{R_{BD}} = \frac{1}{R_1 + R_5} + \frac{1}{R_2} + \frac{1}{R_3 + R_4} = \frac{1}{15\ \Omega} + \frac{1}{6\ \Omega} + \frac{1}{6\ \Omega}$$

$$= (0.0667 + 0.1667 + 0.1667)\frac{1}{\Omega} = (0.400)\frac{1}{\Omega}$$

$$R_{BD} = \frac{1}{0.400}\ \Omega = 2.50\ \Omega$$

Instead of using decimals for the fractions, we could just as well have performed the addition by using the common denominator $6 \times 15 = 90$.

Supplementary Problems

7.17. (a) Find the equivalent resistance of four 60-Ω resistors connected in parallel. (b) If a potential difference of 12 V is applied across the combination, what is the current in each resistor?

7.18. You have three 2-Ω resistors. List the various resistances you can provide with them.

7.19. A 100-Ω resistor and a 200-Ω resistor are connected in parallel with a 40-V power source. (a) What is the current in each resistor? (b) How much power does each one dissipate?

7.20. A 5-Ω resistor is connected in parallel with a 15-Ω resistor. When a potential difference is applied across the combination, which resistor will carry the greater current? What will the ratio of the currents be?

7.21. (a) Find the equivalent resistance of the combination of a 10-Ω resistor connected in parallel with a 30-Ω resistor. (b) A 20-V potential difference is applied across the combination. Find the current in each resistor and the total current.

7.22. (*a*) Find the equivalent resistance of the combination of a 50-Ω, a 75-Ω, and a 100-Ω resistor connected in parallel. (*b*) A 100-V potential difference is applied across the combination. Find the current in each resistor and the total current.

7.23. How can the resistance of a 20-Ω circuit be reduced to an equivalent resistance of 5 Ω?

7.24. How can the resistance of a 1000-Ω circuit be reduced to an equivalent resistance of 200 Ω?

7.25. (*a*) Find the equivalent resistance of the circuit shown in Fig. 7-15. (*b*) If a potential difference of 20 V is applied to the circuit, find the current in each resistor.

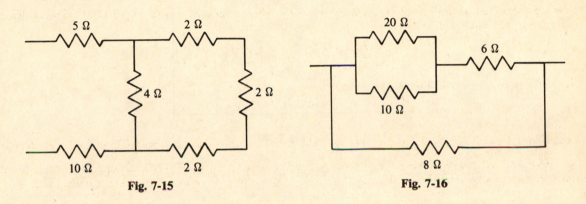

Fig. 7-15   Fig. 7-16

7.26. (*a*) Find the equivalent resistance of the circuit shown in Fig. 7-16. (*b*) If a potential difference of 100 V is applied to the circuit, find the current in each resistor.

7.27. (*a*) Find the equivalent resistance of the circuit shown in Fig. 7-17. (*b*) A 6-V battery whose internal resistance is 1 Ω is connected to the circuit. Find the current in each resistor.

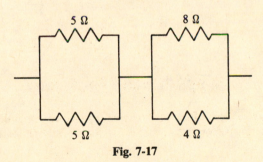

Fig. 7-17

7.28. Two batteries in parallel, each of emf 10 V and internal resistance 0.5 Ω, are connected to an external 20-Ω resistor. Find the current in the external resistor.

7.29. A galvanometer has a resistance of 20 Ω and a range of 0–5 mA. (*a*) What shunt resistance is needed to convert the meter to a 0–100-mA ammeter? (*b*) What series resistance is needed to convert the meter to a 0–10-V voltmeter?

7.30. A 0–0.5-A ammeter has an equivalent resistance of 0.1 Ω. (*a*) If this meter is used as a voltmeter, what range of potential differences can it measure? (*b*) What series resistance is needed for the meter to have a range of 0–1000 V?

7.31. You have a 0–10-mA galvanometer whose resistance is 20 Ω and a separate 20-Ω resistor. What are the maximum ranges of the ammeter and voltmeter than can be assembled from the meter and resistor?

7.32. A 2000-Ω voltmeter reads 10 V when it is in parallel with a resistor of unknown resistance. At the same time an ammeter in series with the combinations reads 0.1 A. Find the unknown resistance.

7.33. Find the equivalent resistance of the circuit of Fig. 7-18.

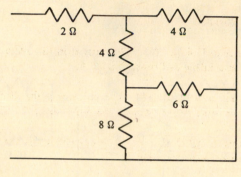

Fig. 7-18

7.34. Find the value of the resistors in the star circuit equivalent to the delta circuit of Fig. 7-19.

7.35. Find the resistances between the points *A* and *B*, *A* and *C*, and *B* and *C* in the circuit of Fig. 7-19.

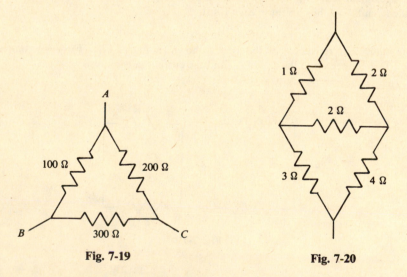

Fig. 7-19   **Fig. 7-20**

7.36. Find the resistance of the bridge circuit of Fig. 7-20.

7.37. A potential difference of 10 V is applied across the bridge circuit of Fig. 7-20. Find the total current and the current in the 3-Ω resistor.

Answers to Supplementary Problems

7.17. (*a*) 15 Ω (*b*) 0.2 A

7.18. 0.67 Ω; 1 Ω; 2 Ω; 3 Ω; 4 Ω; 6 Ω

7.19. (*a*) 0.4 A; 0.2 A   (*b*) 16 W; 8 W

7.20. The 5-Ω resistor will carry a current three times greater than that carried by the 15-Ω resistor.

7.21. (*a*) 7.5 Ω (*b*) 2 A; 0.67 A; 2.67 A

7.22. (*a*) 23.1 Ω (*b*) 2 A; 1.33 A; 1 A; 4.33 A

7.23. Connect a 6.67-Ω resistor in parallel.

7.24. Connect a 250-Ω resistor in parallel.

7.25. (*a*) The equivalent resistance is 17.4 Ω.   (*b*) The current in the 5-Ω and 10-Ω resistors is 1.15 A; in the three 2-Ω resistors, 0.46 A; and in the 4-Ω resistor, 0.69 A.

7.26. (*a*) The equivalent resistance is 4.90 Ω.   (*b*) The current in the 20-Ω resistor is 2.63 A; in the 10-Ω resistor, 5.26 A; in the 6-Ω resistor, 7.89 A; and in the 8-Ω resistor, 12.50 A.

7.27. (*a*) The equivalent resistance is 5.167 Ω.   (*b*) The current in each 5-Ω resistor is 0.486 A; in the 8-Ω resistor, 0.324 A; and in the 4-Ω resistor, 0.649 A.

7.28. 0.494 A

7.29. (*a*) 1.05 Ω (*b*) 1980 Ω

7.30. (*a*) 0–0.05 V   (*b*) 1999.9 Ω

7.31. 0–20 mA; 0–0.4 V

7.32. 105 Ω

7.33. 4.6 Ω

7.34. $R_A = 33.3$ Ω; $R_B = 50$ Ω; $R_C = 100$ Ω

7.35. $R_{AB} = 83.3$ Ω; $R_{AC} = 133.3$ Ω; $R_{BC} = 150$ Ω

7.36. 2.39 Ω

7.37. 4.18 A; 2.45 A

Chapter 8

Simultaneous Equations and Kirchhoff's Rules

SIMULTANEOUS EQUATIONS

Sometimes a problem has two or more unknown quantities, rather than a single one. In order to solve such a problem for the values of these quantities, as many equations are necessary as there are unknown quantities. For instance, the single equation

$$2x + 1 = y$$

is a statement of a relationship between x and y, but it holds for an infinite number of sets of x and y values. However, if the second equation

$$x + 2y = 12$$

must also be obeyed by x and y, then only when $x = 2$ and $y = 5$ will *both* equations be satisfied. The two equations

$$2x + 1 = y$$

$$x + 2y = 12$$

are said to be *simultaneous equations* since together they specify x and y. Simultaneous equations in two or more variables occur in the analysis of complex electric circuits, which is the subject of the latter part of this chapter.

It is worth keeping in mind that not all combinations of simultaneous equations can be solved. An example of such *inconsistent* equations is

$$x + y = 1$$

$$x + y = 2$$

There are no values of x and y for which both these equations are correct. Inconsistent equations in an electrical problem are a sign that the analysis leading up to them is wrong in some way.

SOLVING SIMULTANEOUS EQUATIONS

Three strategies can be used to solve simultaneous equations. These will be discussed at first for problems that involve two unknown quantities, but (as will be seen later) the same techniques can be applied when three or more unknown quantities are present.

The most straightforward method is simply to solve one of the equations for one of the unknowns and then to substitute the expression for this unknown into the other equation.

Problem 8.1. Solve the simultaneous equations

$$x + y = 6$$

$$x - y = 4$$

The first step is to solve $x + y = 6$ for x:

$$x + y = 6$$

$$x = 6 - y$$

Next we substitute this value of x in the second equation and then solve for y:

$$x - y = 4$$
$$6 - y - y = 4$$
$$6 - 2y = 4$$
$$-2y = 4 - 6 = -2$$
$$\frac{-2y}{-2} = \frac{-2}{-2}$$
$$y = 1$$

With the value of y known to be $y = 1$, we can use either of the original equations to find x:

$$x + y = 6 \qquad\qquad x - y = 4$$
$$x + 1 = 6 \qquad\qquad x - 1 = 4$$
$$x = 6 - 1 = 5 \qquad x = 4 + 1 = 5$$

Problem 8.2. Solve the simultaneous equations

$$\frac{x}{4} + 5y = 0$$
$$x - 3y = 4.6$$

From the first equation

$$\frac{x}{4} + 5y = 0$$
$$\frac{x}{4} = -5y$$
$$x = 4 \times (-5y) = -20y$$

Using this value of x in the second equation gives

$$x - 3y = 4.6$$
$$-20y - 3y = 4.6$$
$$-23y = 4.6$$
$$y = -\frac{4.6}{23} = -0.2$$

Since we know that $x = -20y$,

$$x = -20y = -20 \times (-0.2) = 4$$

In some cases it may be easier to begin by solving both equations for one of the unknowns and then set the two resulting expressions equal to each other. This gives one equation in the other unknown which can then be solved in the usual way.

Problem 8.3. Solve the simultaneous equations

$$2x + 1 = y$$
$$x + 2y = 12$$

We begin by solving both equations for y:

$$2x + 1 = y \qquad\qquad x + 2y = 12$$
$$y = 2x + 1 \qquad\qquad 2y = 12 - x$$
$$y = 6 - \frac{x}{2}$$

Now we proceed as follows to find x:

$$y = y$$

$$2x + 1 = 6 - \frac{x}{2}$$

$$2x + \frac{x}{2} = 6 - 1$$

$$\frac{5}{2}x = 5$$

$$x = \frac{2 \times 5}{5} = 2$$

With the value of x known, it is easy to find y:

$$y = 2x + 1 = 2(2) + 1 = 4 + 1 = 5$$

Problem 8.4.   Solve the simultaneous equations

$$\frac{x}{2} - \frac{y}{3} = 2$$

$$\frac{x}{2} + \frac{y}{3} = 0$$

Here it is easiest to solve both equations for $x/2$ and set the results equal:

$$\frac{x}{2} - \frac{y}{3} = 2 \qquad\qquad \frac{x}{2} + \frac{y}{3} = 0$$

$$\frac{x}{2} = 2 + \frac{y}{3} \qquad\qquad \frac{x}{2} = -\frac{y}{3}$$

$$\frac{x}{2} = \frac{x}{2}$$

$$2 + \frac{y}{3} = -\frac{y}{3}$$

Rearranging terms,

$$\frac{2y}{3} = -2$$

$$y = -3$$

From the first equation

$$\frac{x}{2} - \frac{y}{3} = 2$$

$$\frac{x}{2} - \frac{-3}{3} = 2$$

$$\frac{x}{2} + 1 = 2$$

$$\frac{x}{2} = 1$$

$$x = 2$$

The third method of solving simultaneous equations is really just a modification of the second, and is often the least laborious.   The procedure is to multiply both sides of one equation by an appropriate number that will lead to the elimination of one of the unknowns when the two equations are either added or subtracted.   The decision as to whether to add the equations or subtract is determined by which of the operations will eliminate an unknown.   Sometimes it is not even necessary to perform any multiplication first.

Problem 8.5.   Solve the simultaneous equations

$$x + y = 8$$

$$-x + 2y = 1$$

Since in this case adding the two equations will eliminate the x's, the simplest thing to do is just to add the two equations together, which means adding the two left-hand sides and the two right-hand sides separately:

$$
\begin{array}{r}
x + y = 8 \\
[+] \qquad -x + 2y = 1
\end{array}
$$

Adding the equations gives $x + y - x + 2y = 8 + 1$

The x's drop out, as planned $3y = 9$

Solve for y $y = 9/3 = 3$

With the value of y found, we substitute it in either of the original equations to find x:

$$x + y = 8$$
$$x + 3 = 8$$
$$x = 8 - 3 = 5$$

The solution is $y = 3$, $x = 5$.

Problem 8.6. Solve the simultaneous equations

$$3x - y = 1$$
$$x - 2y = 7$$

We can eliminate either x or y to start with. If we wish to eliminate x, we begin by multiplying both sides of the second equation by 3 in order that the coefficients of x be the same in both equations:

$$3(x - 2y) = 3(7)$$
$$3x - 6y = 21$$

Now we can eliminate the x's by subtracting this equation from the first one:

$$
\begin{array}{r}
3x - y = 1 \\
[-] \qquad 3x - 6y = 21
\end{array}
$$

Subtracting the equations gives $3x - y - (3x - 6y) = 1 - (21)$

$$3x - y - 3x + 6y = 1 - 21$$

The x's drop out, as planned $5y = -20$

Solve for y $y = -20/5 = -4$

Finally we substitute $y = -4$ in the first equation to obtain x:

$$3x - y = 1$$
$$3x - (-4) = 1$$
$$3x + 4 = 1$$
$$3x = 1 - 4 = -3$$
$$x = -\frac{3}{3} = -1$$

The solution is $y = -4$, $x = -1$.

MORE THAN TWO UNKNOWNS

As a general rule, the number of simultaneous equations must equal the number of unknown quantities in order to find the values of the unknowns. We have already seen how two simultaneous equations can be solved for two unknowns. The same procedures can be used to solve three or more equations in three or more unknowns, although, of course, more steps are necessary.

Problem 8.7. Solve the following simultaneous equations for x, y, and z:

$$2x + y = 9$$
$$x - y - z = 0$$
$$x + 2y + 3z = 3$$

Let us use the first equation to eliminate y from the other equations. We have

$$2x + y = 9$$
$$y = 9 - 2x$$

$x - y - z = 0$	$x + 2y + 3z = 3$
$x - 9 + 2x - z = 0$	$x + 2(9 - 2x) + 3z = 3$
$3x - z = 9$	$x + 18 - 4x + 3z = 3$
	$-3x + 3z = -15$
	$x - z = 5$

Now we can eliminate the z's by subtracting the last of the right-hand equations from the last of the left-hand equations and solving for x:

$$
\begin{array}{r}
3x - z = 9 \\
[-] \quad x - z = 5 \\
\hline
3x - z - x + z = 9 - 5 \\
2x = 4 \\
x = 2
\end{array}
$$

Hence we have for y

$$y = 9 - 2x = 9 - 2(2) = 9 - 4 = 5$$

and for z

$$x - z = 5$$
$$2 - z = 5$$
$$-z = 5 - 2 = 3$$
$$z = -3$$

Problem 8.8. Solve the following simultaneous equations for x, y, and z:

$$x + y + z = 6$$
$$2x + y - z = 1$$
$$x - 5y - z = -12$$

We begin by using the first equation to eliminate z from the other two:

$$x + y + z = 6$$
$$z = 6 - x - y$$

$2x + y - z = 1$	$x - 5y - z = -12$
$2x + y - (6 - x - y) = 1$	$x - 5y - (6 - x - y) = -12$
$2x + y - 6 + x + y = 1$	$x - 5y - 6 + x + y = -12$
$3x + 2y = 1 + 6 = 7$	$2x - 4y = -12 + 6 = -6$

Now we multiply the last of the left-hand equations by 2 and add the last of the right-hand equations to it, which eliminates y and permits us to find x:

$$
\begin{array}{r}
[\times 2] \quad 3x + 2y = 7 \\
6x + 4y = 14 \\
[+] \quad 2x - 4y = -6 \\
\hline
6x + 4y + 2x - 4y = 14 - 6 \\
8x = 8 \\
x = 1
\end{array}
$$

Using this value of x in the first two original equations permits us to solve them for y and z:

$$x + y + z = 6 \qquad\qquad 2x + y - z = 1$$

$$1 + y + z = 6 \qquad\qquad 2(1) + y - z = 1$$

$$y + z = 6 - 1 = 5 \qquad y - z = 1 - 2 = -1$$

Adding these equations allows us to eliminate z and gives

$$
\begin{aligned}
y + z &= 5 \\
[+] \quad y - z &= -1 \\
\hline
y + z + y - z &= 5 - 1 \\
2y &= 4 \\
y &= 2
\end{aligned}
$$

Since $y + z = 5$,   then $z = 5 - y = 5 - 2 = 3$.

KIRCHHOFF'S RULES

The current that flows in each branch of a complex circuit can be found by applying *Kirchhoff's rules* to the circuit.   The first rule applies to *junctions* of three or more wires (Fig. 8-1).   The second rule applies to *loops*, which are closed conducting paths in the circuit.   The rules are:

1. The sum of the currents that flow into a junction is equal to the sum of the currents that flow out of the junction.

2. The sum of the emf's around a loop is equal to the sum of the IR potential drops around the loop.

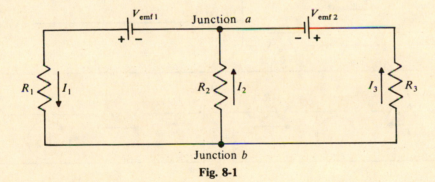

Fig. 8-1

The procedure for applying Kirchhoff's rules is as follows.   First, a direction is arbitrarily chosen for the current in each resistance, as in Fig. 8-1.   If the choice is correct, the current will be found to be positive; if not, the current will be found to be negative, which means that the actual current is in the opposite direction.   Second, in going around a loop (which can be done either clockwise or counterclockwise) an emf is considered positive if the $-$ terminal of its source is met first, negative if the $+$ terminal is met first.   Third, an IR drop is considered positive if the current in the resistance R is in the same direction as the path being followed, negative if the direction is opposite to the path.

In the case of the circuit shown in Fig. 8-1, Kirchhoff's first rule when applied to either junction a or junction b yields

$$I_1 = I_2 + I_3$$

The second rule applied to loop 1, shown in Fig. 8-2(a), and proceeding counterclockwise, yields

$$V_{\text{emf1}} = I_1 R_1 + I_2 R_2$$

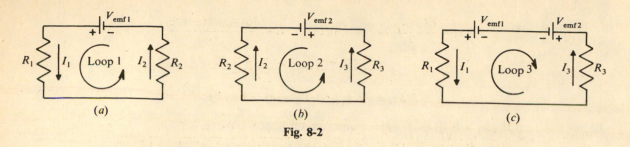

Fig. 8-2

This rule applied to loop 2, shown in Fig. 8-2(b), and again proceeding counterclockwise, yields

$$-V_{emf2} = -I_2R_2 + I_3R_3$$

There is also a third loop, namely the outside one shown in Fig. 8-2(c), which must similarly obey Kirchhoff's second rule.   For the sake of variety we now proceed clockwise and obtain

$$-V_{emf1} + V_{emf2} = -I_3R_3 - I_1R_1$$

Note that this last equation is just the negative sum of the two preceding equations.   Thus, we may use the junction equation and *any two* of the loop equations to solve for the unknown currents I_1, I_2, and I_3.

Problem 8.9.   Two batteries in parallel, one of emf 6 V and internal resistance 0.5 Ω and the other of emf 8 V and internal resistance 0.6 Ω, are connected to an external 10-Ω resistor, as in Fig. 8-3(a). Find the current in the external resistor.

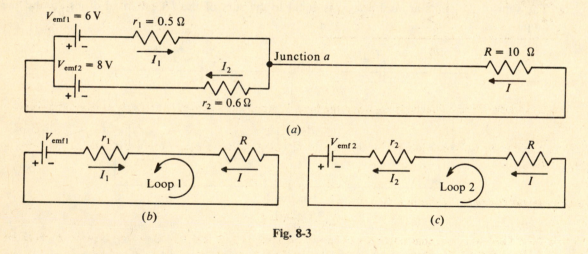

Fig. 8-3

Directions for the currents I, I_1, and I_2 are assumed as in Fig. 8-3(a).   At junction a, Kirchhoff's first law gives

$$I_2 = I + I_1$$

For the loop of Fig. 8-3(b), which we go around counterclockwise,

$$V_{emf1} = IR - I_1r_1$$

and for the loop of Fig. 8-3(c), which we also go around counterclockwise,

$$V_{emf2} = IR + I_2r_2$$

Since $I_2 = I + I_1$,

$$V_{emf2} = IR + Ir_2 + I_1r_2$$

We now solve this equation and the first loop equation for I_1, set the two expressions for I_1 equal, and then solve for I:

$$I_1 = \frac{IR - V_{emf1}}{r_1} \quad \text{and} \quad I_1 = \frac{V_{emf2} - IR - Ir_2}{r_2}$$

$$\frac{IR - V_{emf1}}{r_1} = \frac{V_{emf2} - IR - Ir_2}{r_2}$$

$$I(Rr_2 + Rr_1 + r_2r_1) = V_{emf2}r_1 + V_{emf1}r_2$$

$$I = \frac{V_{emf2}r_1 + V_{emf1}r_2}{Rr_2 + Rr_1 + r_2r_1} = \frac{(8 \text{ V} \times 0.5 \text{ }\Omega) + (6 \text{ V} \times 0.6 \text{ }\Omega)}{(10 \text{ }\Omega \times 0.6 \text{ }\Omega) + (10 \text{ }\Omega \times 0.5 \text{ }\Omega) + (0.6 \text{ }\Omega \times 0.5 \text{ }\Omega)} = 0.673 \text{ A}$$

Problem 8.10. Find the currents in the three resistors of the circuit shown in Fig. 8-4(a). The internal resistances of the emf sources are included in R_1 and R_3.

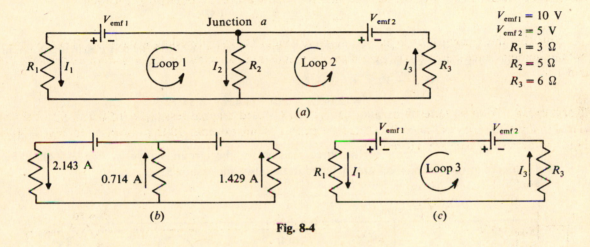

$$
\begin{aligned}
V_{emf1} &= 10 \text{ V} \\
V_{emf2} &= 5 \text{ V} \\
R_1 &= 3 \text{ }\Omega \\
R_2 &= 5 \text{ }\Omega \\
R_3 &= 6 \text{ }\Omega
\end{aligned}
$$

Fig. 8-4

We assume the current directions shown in the figure.   Applying Kirchhoff's first rule to junction a yields

$$I_3 = I_1 + I_2$$

Next we analyze the two inner loops with the help of Kirchhoff's second rule.   Proceeding counterclockwise in loop 1 yields

$$V_{emf1} = I_1R_1 - I_2R_2$$

and proceeding counterclockwise in loop 2 yields

$$V_{emf2} = I_2R_2 + I_3R_3$$

We now have three equations that relate the unknown quantities I_1, I_2, and I_3.   One way to proceed (there are others, equally suitable) is to substitute $I_3 = I_1 + I_2$ in the second loop equation, which gives

$$V_{emf2} = I_2R_2 + I_1R_3 + I_2R_3$$

$$I_1 = \frac{V_{emf2} - I_2R_2 - I_2R_3}{R_3}$$

From the first loop equation,

$$I_1 = \frac{V_{emf1} + I_2R_2}{R_1}$$

Since these two expressions must be equal,

$$\frac{V_{emf1} + I_2R_2}{R_1} = \frac{V_{emf2} - I_2R_2 - I_2R_3}{R_3}$$

At this point we substitute the values of the various emf's and resistances and solve for I_2.   This substitution can also be done earlier or later in a calculation of this kind, whatever seems most convenient.   The calculation proceeds as follows:

$$\frac{10\,V + (5\,\Omega)I_2}{3\,\Omega} = \frac{5\,V - (5\,\Omega)I_2 - (6\,\Omega)I_2}{6\,\Omega}$$

$$\frac{10\,V}{3\,\Omega} + \left(\frac{5\,\Omega}{3\,\Omega}\right)I_2 = \frac{5\,V}{6\,\Omega} - \left(\frac{5\,\Omega}{6\,\Omega}\right)I_2 - \left(\frac{6\,\Omega}{6\,\Omega}\right)I_2$$

$$\left(\frac{5}{3} + \frac{5}{6} + \frac{6}{6}\right)I_2 = \left(\frac{5}{6} - \frac{10}{3}\right)A$$

$$I_2 = -0.714\,A$$

The minus sign means that the current I_2 is in the opposite direction to the one shown in the figure. From the first loop equation,

$$I_1 = \frac{V_{emf1} + I_2 R_2}{R_1} = \frac{10\,V - (0.714\,A \times 5\,\Omega)}{3\,\Omega} = 2.143\,A$$

Finally we find I_3 from the junction equation:

$$I_3 = I_1 + I_2 = 2.143\,A - 0.714\,A = 1.429\,A$$

The actual currents are shown in Fig. 8-4(*b*).

As a check on the calculation we can apply Kirchhoff's second rule to the outside loop of the circuit, shown in Fig. 8-4(*c*). Proceeding counterclockwise,

$$V_{emf2} + V_{emf1} = I_1 R_1 + I_3 R_3$$

$$5\,V + 10\,V = (2.143\,A \times 3\,\Omega) + (1.429\,A \times 6\,\Omega)$$

$$15\,V = 15\,V$$

Problem 8.11. Find the currents in the three resistors of the circuit shown in Fig. 8-5(*a*). The internal resistances of the emf sources are included in the resistances shown.

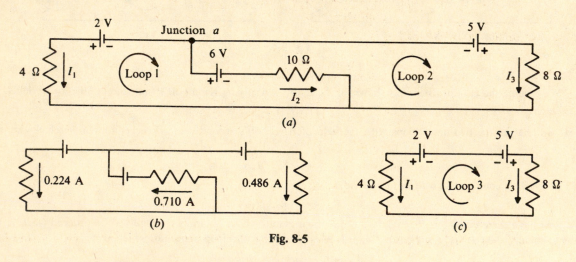

Fig. 8-5

Let us now work directly from the numerical values of the emf's and resistances in the figure. Applying Kirchhoff's first rule to junction *a*, assuming the current directions shown, yields

$$I_1 + I_2 + I_3 = 0$$

Clearly one or two current directions are incorrect, but this makes no difference since the result will be a negative current in those cases. Next we apply Kirchhoff's second rule to loop 1 and proceed clockwise:

$$-2\,V - 6\,V = (10\,\Omega)I_2 - (4\,\Omega)I_1$$

The emf's are considered negative because we encountered their + terminals first. Solving for I_2 gives

$$I_2 = \frac{-8\,V}{10\,\Omega} + \left(\frac{4\,\Omega}{10\,\Omega}\right)I_1 = -0.8\,A + 0.4I_1$$

Now we proceed clockwise in loop 2 and obtain

$$6\,V + 5\,V = (8\,\Omega)I_3 - (10\,\Omega)I_2$$

Substituting $I_3 = -I_1 - I_2$ and solving for I_2 yields

$$11\text{ V} = -(8\,\Omega)I_1 - (8\,\Omega)I_2 - (10\,\Omega)I_2$$

$$I_2 = \frac{-11\text{ V}}{18\,\Omega} - \left(\frac{8\,\Omega}{18\,\Omega}\right)I_1 = -0.611\text{ A} - 0.444I_1$$

Setting equal the two expressions for I_2 and solving for I_1,

$$-0.8\text{ A} + 0.4I_1 = -0.611\text{ A} - 0.444I_1$$

$$0.844I_1 = 0.189\text{ A}$$

$$I_1 = 0.224\text{ A}$$

From the first loop equation

$$I_2 = -0.8\text{ A} + 0.4I_1 = -0.710\text{ A}$$

and so

$$I_3 = -I_1 - I_2 = -0.224\text{ A} + 0.710\text{ A} = 0.486\text{ A}$$

The actual currents are shown in Fig. 8-5(b).

Again we check the results by using the outside loop of the circuit as shown in Fig. 8-5(c). Proceeding clockwise,

$$-2\text{ V} + 5\text{ V} = (8\,\Omega \times 0.486\text{ A}) - (4\,\Omega \times 0.224\text{ A})$$

$$3\text{ V} = 3\text{ V}$$

THE L ATTENUATOR

When a source of electric energy is connected to a load, the power transfer is a maximum when both source and load have the same resistance. For a specific example, let us consider a 100-V battery, whose internal resistance is $10\,\Omega$, that is connected to a load resistor R, as in Fig. 8-6. The total resistance in the circuit is $R + r$ so the current in R is

$$I = \frac{\text{battery emf}}{\text{total resistance}} = \frac{V_{\text{emf}}}{R + r} = \frac{100\text{ V}}{R + 10\,\Omega}$$

The power dissipated in the load R is

$$P = I^2 R$$

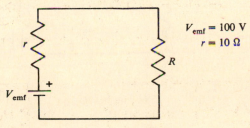

$$V_{\text{emf}} = 100\text{ V}$$
$$r = 10\,\Omega$$

Fig. 8-6

Here is a table showing how I and P vary with the value of the load resistance R for different values of R:

Load Resistance, R (Ω)	Current, I (A)	Load Power, $P = I^2R$ (W)
2	8.3	139
4	7.1	204
6	6.3	234
8	5.6	247
10	6.0	250
12	4.5	248
14	4.2	243
16	3.8	237
18	3.6	230
20	3.3	222

Clearly the maximum load power occurs when $R = r = 10\,\Omega$.

An *attenuator* is a system of resistors placed between a source and a load that reduces the voltage to the load but keeps the source current unchanged. Thus an attenuator can be used to match the resistance of an external circuit (load plus attenator) to the internal resistance of a source and so maximize the power transfer. The resistors that make up an attenuator are sometimes variable and mounted on the same shaft ("ganged") in such a way that the resistance presented to the source remains constant while the load voltage is varied. An attenuator with fixed resistors is usually called a *pad*.

In the circuit of Fig. 8-7(a), a load resistance of 10 Ω means a current in both source and load of 5 A. The voltage across both source and load is $V = IR = 5\,A \times 10\,\Omega = 50\,V$. Now suppose we want to reduce the voltage across the load to, say, 20 V while keeping the source current unchanged at 5 A. The new load current will be $I_L = V_L/R_L = (20\,V)/(10\,\Omega) = 2\,A$. If we simply insert a series resistor in the circuit, the load voltage will be reduced but so will the current drawn from the source. What we can do instead is use the pair of resistors R_1 and R_2 of Fig. 8-7(b), which make up an L attenuator pad (so-called because their combination is like an inverted letter "L"). Resistor R_1 reduces the voltage supplied to the load, which decreases the load current, while R_2 provides an additional path for the source current, so the source current can remain the same while the load voltage and current are decreased. Kirchhoff's rules can be used to find the values of R_1 and R_2 in a particular case.

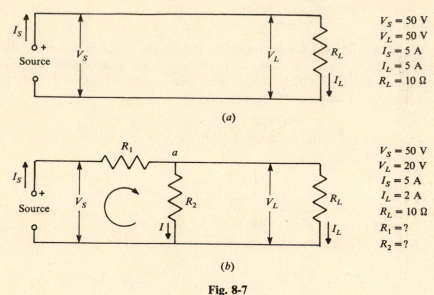

(a)

(b)

Fig. 8-7

Problem 8.12. Find the values of R_1 and R_2 in the L pad of Fig. 8-7(b).

Applying Kirchhoff's first rule to junction a,

$$I_S = I + I_L$$
$$I = I_S - I_L = 5\,A - 2\,A = 3\,A$$

This is the current in the resistor R_2. Since the voltage across R_2 is $V_L = 20\,V$,

$$R_2 = \frac{V_L}{I} = \frac{20\,V}{3\,A} = 6.67\,\Omega$$

To find R_1, we apply Kirchhoff's second rule to the left-hand loop, proceeding clockwise:

$$V_S = I_S R_1 + I R_2$$

Solving for R_1,

$$R_1 = \frac{V_S - I R_2}{I_S} = \frac{V_S - V_L}{I_S} = \frac{50\,V - 20\,V}{5\,A} = 6\,\Omega$$

To check these values of R_1 and R_2, we note that the equivalent resistance R of the combination of R_L, R_1, and R_2

must be the same as the 10-Ω resistance of R_L by itself.   Since R_L and R_2 are in parallel, their equivalent resistance is

$$R' = \frac{R_L \times R_2}{R_L + R_2} = \frac{10\,\Omega \times 6.67\,\Omega}{10\,\Omega + 6.67\,\Omega} = 4\,\Omega$$

This resistance is in series with $R_1 = 6\,\Omega$,   so the equivalent resistance of the combination of R_L, R_1, and R_2 is

$$R = R_1 + R' = 6\,\Omega + 4\,\Omega = 10\,\Omega$$

as it should be.

Problem 8.13.   A 24-V, 4-Ω source is to be connected to a 4-Ω load.   Design an L pad that will give a load current of 2 A without changing the 4-Ω resistance presented to the source.

Without the pad the source current would be

$$I_S = \frac{V_S}{R} = \frac{V_S}{R_S + R_L} = \frac{24\,\text{V}}{4+4} = 3\,\text{A}$$

and the net source voltage would be

$$V_S = V_{\text{emf}} - I_S R_S = 24\,\text{V} - (3\,\text{A} \times 4\,\Omega) = 12\,\text{V}$$

The voltage across the load to give a load current of $I_L = 2\,\text{A}$ must be

$$V_L = I_L R_L = 2\,\text{A} \times 4\,\Omega = 8\,\text{V}$$

We now proceed exactly as in Problem 8.12 with the help of Fig. 8-7(b):

$$I = I_S - I_L = 3\,\text{A} - 2\,\text{A} = 1\,\text{A} \qquad R_2 = \frac{V_L}{I} = \frac{8\,\text{V}}{1\,\text{A}} = 8\,\Omega \qquad R_1 = \frac{V_S - V_L}{I_S} = \frac{12\,\text{V} - 8\,\text{V}}{3\,\text{A}} = 1.33\,\Omega$$

As a check,

$$R = R_1 + \frac{R_L \times R_2}{R_L + R_2} = 1.33\,\Omega + \frac{4\,\Omega \times 8\,\Omega}{4\,\Omega + 8\,\Omega} = 4\,\Omega$$

THE T ATTENUATOR

The T attenuator pad consists of three resistors, R_1, R_2, and R_3, arranged as in Fig. 8-8, with $R_1 = R_3$.   Such an attenuator presents the same resistance to both source and load, unlike an L attenuator.

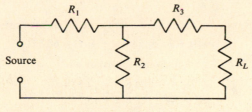

Fig. 8-8

Problem 8.14.   A 120-V, 40-Ω source is connected to a 40-Ω load, as in Fig. 8-9(a).   Design a T pad that will reduce the voltage across the load to 10 V.

The current in the original circuit of Fig. 8-9(a) is

$$I_S = \frac{V_{\text{emf}}}{R_L + r} = \frac{120\,\text{V}}{80\,\Omega} = 1.5\,\text{A}$$

and this must be the source current in the new circuit of Fig. 8-9(b).   With the pad in place, the load current is

$$I_L = \frac{V_L}{R_L} = \frac{10\,\text{V}}{40\,\Omega} = 0.25\,\text{A}$$

Applying Kirchhoff's first rule to junction a gives

$$I_S = I + I_L$$

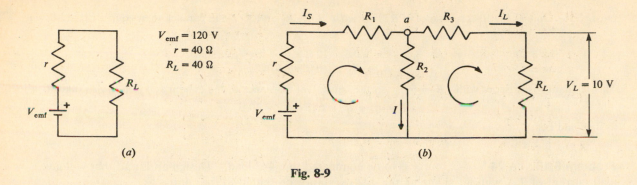

Fig. 8-9

Rearranging terms,

$$I = I_S - I_L = 1.5 \text{ A} - 0.25 \text{ A} = 1.25 \text{ A}$$

Next Kirchhoff's second rule is applied to the two loops of Fig. 8-9(b). Proceeding clockwise,

$$V_{emf} = I_S r + I_S R_1 + I R_2$$
$$0 = I_L R_3 + I_L R_L - I R_2$$

Inserting the values of the known quantities in these equations and substituting R_1 for R_3 (which we can do since $R_1 = R_3$) yields

$$120 \text{ V} = (1.5 \text{ A} \times 40 \text{ Ω}) + (1.5 \text{ A})R_1 + (1.25 \text{ A})R_2$$
$$0 = (0.25 \text{ A})R_1 + (0.25 \text{ A} \times 40 \text{ Ω}) - (1.25 \text{ A})R_2$$

We now have two equations in the two unknowns R_1 and R_2. We can readily eliminate R_2 by adding the two equations together. Doing this and performing the indicated multiplications enables us to find R_1 and R_3:

$$120 \text{ V} = 60 \text{ V} + (1.5 \text{ A})R_1 + (0.25 \text{ A})R_1 + 10 \text{ V}$$
$$(1.75 \text{ A})R_1 = 50 \text{ V}$$
$$R_1 = \frac{50 \text{ V}}{1.75 \text{ A}} = 28.6 \text{ Ω}$$
$$R_3 = R_1 = 28.6 \text{ Ω}$$

From the second loop equation,

$$(1.25 \text{ A})R_2 = (0.25 \text{ A})R_1 + (0.25 \text{ A} \times 40 \text{ Ω}) = 7.1 \text{ V} + 10 \text{ V} = 17.1 \text{ V}$$
$$R_2 = \frac{17.1 \text{ V}}{1.25 \text{ A}} = 13.7 \text{ Ω}$$

To check, we note first that R_3 and R_L are in series with an equivalent resistance R' of

$$R' = R_3 + R_L = 28.6 \text{ Ω} + 40 \text{ Ω} = 68.8 \text{ Ω}$$

This combination is in parallel with R_2 for an equivalent resistance of

$$R'' = \frac{R' \times R_2}{R' + R_2} = \frac{68.6 \text{ Ω} \times 13.7 \text{ Ω}}{68.6 \text{ Ω} + 13.7 \text{ Ω}} = 11.4 \text{ Ω}$$

Hence the equivalent resistance of the entire circuit of pad plus load is

$$R = R'' + R_1 = 11.4 \text{ Ω} + 28.6 \text{ Ω} = 40 \text{ Ω}$$

which is the same as the load resistance itself. Hence our work is correct.

Supplementary Problems

8.15. Solve the following simultaneous equations:

(a) $x = 2y$
     $x + 3y = 10$

(b) $x + y = 8$
     $x - 2y = 2$

(c) $x + 2y = 3$
     $2x - 2y = 3$

(d) $2x + 3y = 2$
     $6x - 3y = 2$

(e) $2x + 3y = -8$
     $3x - 5y = 7$

(f) $\dfrac{x}{4} + \dfrac{y}{3} = 2$
     $2x + y = 11$

(g) $\dfrac{x}{2} + y = 0$
     $3x - 4y = 5$

(h) $x + y = 20$
     $\dfrac{x}{2} + 2y = 25$

(i) $3x - y = 0$
     $4x - y = 0.5$

(j) $7x + 3y = 8.5$
     $x - 2y = -1.7$

(k) $x + \dfrac{y}{3} = 1.1$
     $6x + y = 0.3$

(l) $x - 3y = 8$
     $\dfrac{x}{11} + \dfrac{y}{3} = 8$

(m) $x + 6y = 10$
     $3x - 8y = 17$

(n) $x + 2y - z = 7$
     $x + 4y + z = 10$
     $2x - 2y - 2z = 11$

(o) $2x + y = -4$
     $x - y - z = 4$
     $x - 4y + z = 4$

(p) $x + 2y - z = 0$
     $\dfrac{x}{2} + y + 2z = 3$
     $3x + 10y + 5z = 14$

(q) $x + y = 0$
     $2x + y + z = 1.2$
     $x - y - 3z = -5.1$

(r) $x - y - 2z = 20$
     $x + \dfrac{y}{2} + z = 20$
     $2x + 2y + 5z = 15$

(s) $y - 2z = 1.4$
     $2x + y + z = -4.6$
     $2x - 1.2y + 3z = 0$

(t) $x + 2y + 3z = -16$
     $2x - y - 4z = 3$
     $\dfrac{x}{2} - 3y + z = 5$

8.16. Find the currents in the resistors of the circuit shown in Fig. 8-10. The internal resistances of the emf sources are included in the external resistances.

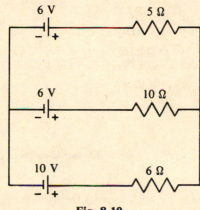

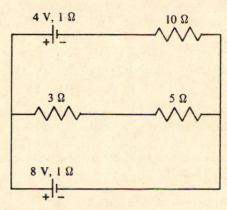

Fig. 8-10

Fig. 8-11

8.17. Find the currents in the resistors of the circuit shown in Fig. 8-11. The internal resistances of the emf sources must be taken into account.

8.18. Find the currents in the resistors of the circuit shown in Fig. 8-12. The internal resistances of the emf sources are included in the external resistances.

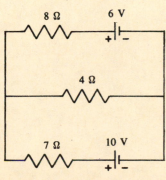

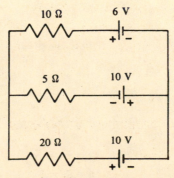

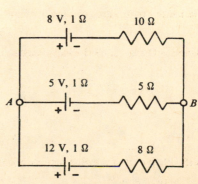

Fig. 8-12

Fig. 8-13

Fig. 8-14

8.19. Find the currents in the resistors of the circuit shown in Fig. 8-13. The internal resistances of the emf sources are included in the external resistances.

8.20. (a) Find the current in the 5-Ω resistor in the circuit shown in Fig. 8.14. (b) Find the potential difference between points A and B. The internal resistances of the emf sources must be taken into account.

8.21. The voltage across a 3-Ω load connected to a source is 6 V. Design an L pad that will reduce the voltage across the load to 2 V while leaving the source current unchanged.

8.22. The voltage across an 80-Ω load connected to a source is 100 V. Design an L pad that will reduce the voltage across the load to 25 V while leaving the source current unchanged.

8.23. A 40-V, 5-Ω source is connected to a 5-Ω load. Design a T pad that will reduce the voltage across the load to 15 V.

Answers to Supplementary Problems

8.15.

(a)	$x = 4, y = 2$	(h)	$x = 10, y = 10$	(o)	$x = -1, y = -2, z = -3$
(b)	$x = 6, y = 2$	(i)	$x = 0.5, y = 1.5$	(p)	$x = -1, y = 1.1, z = 1.2$
(c)	$x = 2, y = 1/2$	(j)	$x = 0.7, y = 1.2$	(q)	$x = -0.3, y = 0.3, z = 1.5$
(d)	$x = 1/2, y = 1/3$	(k)	$x = -1, y = 6.3$	(r)	$x = 20, y = 50, z = -25$
(e)	$x = -1, y = -2$	(l)	$x = 44, y = 12$	(s)	$x = 1.8, y = -5, z = -3.2$
(f)	$x = 4, y = 3$	(m)	$x = 7, y = 1/2$	(t)	$x = -4, y = -3, z = -2$
(g)	$x = 1, y = -1/2$	(n)	$x = 7, y = 1/2, z = 1$		

8.16. The current in the 5-Ω resistor is 0.286 A to the left, the current in the 10-Ω resistor is 0.143 A to the left, and the current in the 6-Ω resistor is 0.429 A to the right.

8.17. The current in the 10-Ω resistor is 0.935 A to the left and the currents in the 3-Ω and 5-Ω resistors are both 0.785 A to the left.

8.18. The current in the 8-Ω resistor is 0.22 A to the left, the current in the 4-Ω resistor is 1.05 A to the right, and the current in the 7-Ω resistor is 0.83 A to the left.

8.19. The current in the 10-Ω resistor is 0.857 A to the left, the current in the 5-Ω resistor is 1.486 A to the right, and the current in the 20-Ω resistor is 0.629 A to the left.

8.20. (a) 0.475 A to the right   (b) 7.85 V   **8.22.** $R_1 = 60\ \Omega, R_2 = 26.67\ \Omega$

8.21. $R_1 = 2\ \Omega, R_2 = 1.5\ \Omega$ **8.23.** $R_1 = R_3 = 0.714\ \Omega, R_2 = 17.143\ \Omega$

Chapter 9

Inductance

SELF-INDUCTANCE

When the current in a circuit changes, the magnetic field enclosed by the circuit also changes, and the resulting change in flux leads to a *self-induced emf* of

$$\text{Self-induced emf} = V = -L\frac{\Delta I}{\Delta t}$$

Here $\Delta I/\Delta t$ is the rate of change of the current and L is a property of the circuit called its *self-inductance*, or, more commonly, its *inductance*. The minus sign indicates that the direction of V is such as to oppose the change in current ΔI that caused it.

The unit of inductance is the *henry* (H). A circuit or circuit element (such as a solenoid) that has an inductance of 1 H will have a self-induced emf of 1 V when the current through it changes at the rate of 1 A/s. Because the henry is a rather large unit, the *millihenry* and *microhenry* are often used, where

$$1 \text{ millihenry} = 1 \text{ mH} = 10^{-3} \text{ H} \qquad 1 \text{ microhenry} = 1 \ \mu\text{H} = 10^{-6} \text{ H}$$

A *solenoid* is a coil in the form of a helix, as in Fig. 9-1. The inductance of a solenoid is

$$L = \frac{\mu N^2 A}{l}$$

where μ is the magnetic permeability of the core material, N is the number of turns, A is the cross-sectional area, and l is the length of the solenoid. In free space, $\mu_0 = 1.26 \times 10^{-6}$ H/m; the value of μ in air is very nearly the same.

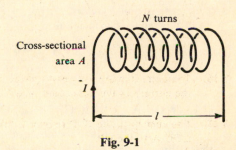

Fig. 9-1

Problem 9.1. Find the average self-induced emf in a 25-mH solenoid when the current in it falls from 0.2 A to 0 in 0.01 s.

Since all we want is the magnitude of V, we can disregard the minus sign, and

$$V = L\frac{\Delta I}{\Delta t} = 25 \times 10^{-3} \text{ H} \times \frac{0.2 \text{ A}}{0.01 \text{ s}} = 0.5 \text{ V}$$

Problem 9.2. The current in a circuit falls from 5 A to 1 A in 0.1 s. If an average emf of 2 V is induced in the circuit while this is happening, find the inductance of the circuit.

Since $V = -L\,\Delta I/\Delta t$, we have (disregarding the minus sign)

$$L = \frac{V\,\Delta t}{\Delta I} = \frac{2 \text{ V} \times 0.1 \text{ s}}{5 \text{ A} - 1 \text{ A}} = 0.05 \text{ H}$$

Problem 9.3. Find the inductance in air of a 500-turn solenoid 10 cm long whose cross-sectional area is 20 cm^2.

First we must express l and A in m and m^2, respectively. Since $1 \text{ cm} = 10^{-2} \text{ m}$ and $1 \text{ cm}^2 = 10^{-4} \text{ m}^2$,

$$l = 10 \text{ cm} = 10 \times 10^{-2} \text{ m} = 10^{-1} \text{ m} \qquad A = 20 \text{ cm}^2 = 20 \times 10^{-4} \text{ m}^2 = 2 \times 10^{-3} \text{ m}^2$$

With $\mu = \mu_0$, the inductance is

$$L = \frac{\mu_0 N^2 A}{l} = \frac{1.26 \times 10^{-6} \text{ H/m} \times (500)^2 \times 2 \times 10^{-3} \text{ m}^2}{10^{-1} \text{ m}} = 6.3 \times 10^{-3} \text{ H} = 6.3 \text{ mH}$$

Problem 9.4. A solenoid 20 cm long and 2 cm in diameter has an inductance of 0.178 mH. How many turns of wire does it have?

The radius of the solenoid is $r = 1 \text{ cm} = 0.01 \text{ m}$ and so its cross-sectional area is

$$A = \pi r^2 = \pi \times (0.01 \text{ m})^2 = 3.14 \times 10^{-4} \text{ m}^2$$

Since $L = \mu_0 N^2 A / l$ in air,

$$N = \sqrt{\frac{Ll}{\mu_0 A}} = \sqrt{\frac{0.178 \times 10^{-3} \text{ H} \times 0.2 \text{ m}}{1.26 \times 10^{-6} \text{ H/m} \times 3.14 \times 10^{-4} \text{ m}^2}} = 300 \text{ turns}$$

Problem 9.5. An inductor consists of an iron ring 5 cm in diameter and 1 cm^2 in cross-sectional area that is wound with 1000 turns of wire. If the permeability of the iron is constant at 400 times that of free space at the magnetic fields at which the inductor will be used, find its inductance.

An inductor of this kind is essentially a solenoid bent into a circle. The length of the equivalent solenoid is therefore

$$l = \pi D = \pi \times 0.05 \text{ m} = 0.157 \text{ m}$$

and its cross-sectional area is $A = 1 \text{ cm}^2 = 10^{-4} \text{ m}^2$. The permeability of the core is $\mu = 400\mu_0$. The inductance of this inductor is therefore

$$L = \frac{\mu N^2 A}{l} = \frac{400 \times 1.26 \times 10^{-6} \text{ H/m} \times (10^3)^2 \times 10^{-4} \text{ m}^2}{0.157 \text{ m}} = 0.32 \text{ H}$$

ENERGY OF A CURRENT-CARRYING INDUCTOR

Because a self-induced emf opposes any change in current in an inductor, work has to be done against this emf in order to establish a current in the inductor. This work is stored as magnetic potential energy. If L is the inductance of an inductor, its potential energy when it carries the current I is

$$W = \frac{1}{2} L I^2$$

It is this energy which powers the self-induced emf that opposes any decrease in the current through the inductor.

Problem 9.6. How much magnetic potential energy is stored in a 20-mH coil when it carries a current of 0.2 A?

$$W = \frac{1}{2} L I^2 = \frac{1}{2} \times 20 \times 10^{-3} \text{ H} \times (0.2 \text{ A})^2 = 4 \times 10^{-4} \text{ J}$$

Problem 9.7. What should the current in a 20-mH coil be in order that it contain 1 J of energy?

$$W = \frac{1}{2} L I^2 = \frac{L I^2}{2}$$

$$L I^2 = 2W$$

$$I^2 = \frac{2W}{L}$$

$$I = \sqrt{\frac{2W}{L}} = \sqrt{\frac{2 \times 1 \text{ J}}{20 \times 10^{-3} \text{ H}}} = 10 \text{ A}$$

TIME CONSTANT

Because the self-induced emf in a circuit such as that of Fig. 9-2 is always such as to oppose any changes in the current in the circuit, the current does not rise instantly to its final value of $I = V/R$ when the switch is closed. When the switch in Fig. 9-2 is closed, the current starts to build up, and as a result a self-induced emf, $-L(\Delta I/\Delta t)$, occurs that opposes the battery voltage V. The net voltage in the circuit is therefore $V - L(\Delta I/\Delta t)$, which must equal IR by Ohm's law:

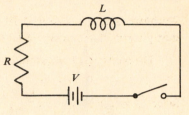

$$V - L\frac{\Delta I}{\Delta t} = IR$$

Impressed voltage − induced emf = net voltage

Fig. 9-2

The rate of increase of the current at the moment when the current is I is accordingly

$$\frac{\Delta I}{\Delta t} = \frac{V - IR}{L}$$

The larger the inductance L, the more gradually the current increases. When the switch is first closed, $I = 0$ and

$$\frac{\Delta I}{\Delta t} = \frac{V}{L}$$

Eventually the current stops rising and $\Delta I/\Delta t = 0$. From then on

$$0 = \frac{V - IR}{L}$$

Multiplying both sides of this equation by L gives $0 = V - IR$ since $0 \times L = 0$, and the final current is

$$I = \frac{V}{R}$$

The final current depends only upon V and R; the effect of L is to delay the establishment of the final current.

As shown in Fig. 9-3, the current in the circuit of Fig. 9-2 rises gradually in such a manner that after a time t equal to L/R it reaches 63 percent of its final value. If the battery is short-circuited, the current decreases in such a manner that after $t = L/R$ it has fallen to 37 percent of its original value (Fig. 9-4). The quantity L/R is called the *time constant T* of the circuit; the smaller the time constant, the more rapidly the current changes.

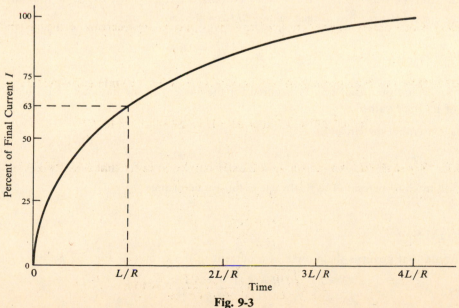

Fig. 9-3

What is happening during the establishment of the current of Fig. 9-3 is that magnetic energy $(1/2)LI^2$ is being absorbed by the inductance L of the circuit. When the battery is short-circuited so that no voltage is impressed on the circuit, the stored energy is what powers the subsequent decreasing current of Fig. 9-4.

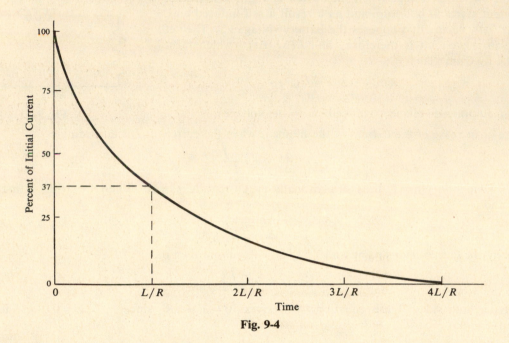

Fig. 9-4

Problem 9.8. A 0.1-H inductor whose resistance is 20 Ω is connected to a 12-V battery of negligible internal resistance. (*a*) What is the initial rate at which the current increases? (*b*) What happens to the rate of current increase? (*c*) What is the final current?

(*a*) At the moment the connection is made,

$$\frac{\Delta I}{\Delta t} = \frac{V_{emf}}{L} = \frac{12 \text{ V}}{0.1 \text{ H}} = 120 \text{ A/s}$$

The initial rate at which the current increases is 120 A/s.

(*b*) Since $\Delta I/\Delta t = (V - IR)/L$, as the current increases, its rate of change $\Delta I/\Delta t$ decreases.

(*c*) When the current has reached its final value, $\Delta I/\Delta t = 0$ and

$$I = \frac{V}{R} = \frac{12 \text{ V}}{20 \text{ Ω}} = 0.6 \text{ A}$$

Problem 9.9. What period of time is required for the current in the inductor of Problem 9.8 to reach 63 percent of its final value?

The time constant of the circuit is

$$T = \frac{L}{R} = \frac{0.1 \text{ H}}{20 \text{ Ω}} = 0.005 \text{ s}$$

The current will reach 63 percent of its final value in this period of time.

EXPONENTIALS

The formula that governs the growth of a current in the circuit of Fig. 9-2 is

$$I = I_0(1 - e^{-t/T})$$

In this formula, which is what is plotted in Fig. 9-3, I_0 is the steady-state current V/R, T is the time constant L/R, and I is the current at the time t after the switch is closed.   The quantity e has the value

$$e = 2.718 \cdots$$

and is often found in equations in engineering and science.   A quantity that consists of e raised to a power is called an *exponential*.   To find the value of e^x or e^{-x}, an electronic calculator or a suitable table can be used.

From the above formula we can see why I reaches 63 percent of its final value $I_0 = V/R$ in the time L/R.   When $t = L/R$, $t/T = 1$,   and

$$I = I_0(1 - e^{-1}) = I_0\left(1 - \frac{1}{e}\right) = I_0(1 - 0.37) = 0.63 I_0$$

In the case of a circuit containing L and R in which the battery is short-circuited,

$$I = I_0 e^{-t/T}$$

Now when $t = L/R$, $t/T = 1$,   and

$$I = I_0 e^{-1} = I_0\left(\frac{1}{e}\right) = 0.37 I_0$$

The current drops to 37 percent of its original value of $I_0 = V/R$ after a time equal to L/R.

Problem 9.10.   A 0.2-H inductor with a resistance of 3 Ω is connected to a 6-V battery whose internal resistance is 1 Ω.   (a) Find the final current in the circuit.   (b) Find the current in the circuit 0.01 s, 0.05 s, and 0.1 s after the connection is made.

(a)   The total resistance in the circuit is the sum of the 3-Ω resistance of the inductor and the 1-Ω internal resistance of the battery, so that

$$R = 3\,\Omega + 1\,\Omega = 4\,\Omega$$

The final current in the circuit is therefore

$$I_0 = \frac{V}{R} = \frac{6\,\text{V}}{4\,\Omega} = 1.5\,\text{A}$$

(b)   The time constant of the circuit is

$$T = \frac{L}{R} = \frac{0.2\,\text{H}}{4\,\Omega} = 0.05\,\text{s}$$

At the time t after the connection is made, the current is given by

$$I = I_0(1 - e^{-t/T})$$

When $t = 0.01$ s, $t/T = (0.01\,\text{s})/(0.05\,\text{s}) = 0.2$.   With the help of a calculator or a table of exponentials, we find that

$$e^{-t/T} = e^{-0.2} = 0.82$$

and so

$$I = I_0(1 - e^{-t/T}) = 1.5\,\text{A}(1 - 0.82) = 1.5\,\text{A} \times 0.18 = 0.27\,\text{A}$$

When $t = 0.05$ s, $t = T$ and

$$I = 0.63 I_0 = 0.63 \times 1.5\,\text{A} = 0.95\,\text{A}$$

When $t = 0.1$ s, $t/T = (0.1\,\text{s})/(0.05\,\text{s}) = 2$,   and

$$e^{-t/T} = e^{-2} = 0.14$$

and so

$$I = I_0(1 - e^{-t/T}) = 1.5\,\text{A}(1 - 0.14) = 1.5\,\text{A} \times 0.86 = 1.29\,\text{A}$$

Problem 9.11.   Find the current in the above inductor 0.01 s and 0.1 s after it has been short-circuited after having been connected to the battery for a long time.

The resistance of the circuit is now just the 3-Ω resistance of the inductor itself.   Hence the time constant is

$$T = \frac{L}{R} = \frac{0.2 \text{ H}}{3 \ \Omega} = 0.067 \text{ s}$$

When $t = 0.01$ s, $t/T = (0.01 \text{ s})/(0.067 \text{ s}) = 0.15$,   and

$$e^{-t/T} = e^{-0.15} = 0.86$$

Hence the current in the inductor is

$$I = I_0 e^{-t/T} = 1.5 \text{ A} \times 0.86 = 1.29 \text{ A}$$

When $t = 0.1$ s, $t/T = (0.1 \text{ s})/(0.067 \text{ s}) = 1.5$,   and

$$e^{-t/T} = e^{-1.5} = 0.22$$

Hence the current in the inductor at this time is

$$I = I_0 e^{-t/T} = 1.5 \text{ A} \times 0.22 = 0.33 \text{ A}$$

INDUCTORS IN COMBINATION

Connecting coils in parallel reduces the total inductance to less than that of any of the individual coils.   This is rarely done because coils are relatively large and expensive compared with other electronic components; a coil of the required smaller inductance would normally be used in the first place.   On the other hand, coils are sometimes connected in series, in which case the inductance of the combination is the sum of the inductances of the individual coils:

$$L = L_1 + L_2 + L_3 + \cdots \qquad \text{(inductors in series)}$$

Because the magnetic field of a current-carrying coil extends beyond the inductor itself, the total inductance of two or more coils connected together will be changed if they are close to one another. Depending on how the coils are arranged, the total inductance may be larger or smaller than if the coils were farther apart.   This effect is called *mutual inductance* and is not considered in the above formula.

Problem 9.12.   The 0.2-H, 3-Ω inductor of Problem 9.10 is connected in series with another 0.2-H, 3-Ω inductor and the combination is again connected to a 6-V battery of internal resistance 1 Ω.   (a) Find the time constant of the new circuit and the final current.   (b) If the battery is short-circuited so the current flows only through the inductors until it dies away, find the new time constant.

(a)   The total inductance and total resistance are respectively

$$L = L_1 + L_2 = 0.2 \text{ H} + 0.2 \text{ H} = 0.4 \text{ H} \qquad R = R_1 + R_2 + R_3 = 3 \ \Omega + 3 \ \Omega + 1 \ \Omega = 7 \ \Omega$$

The time constant is therefore

$$T = \frac{L}{R} = \frac{0.4 \text{ H}}{7 \ \Omega} = 0.057 \text{ s}$$

The final current is

$$I_0 = \frac{V}{R} = \frac{6 \text{ V}}{7 \ \Omega} = 0.86 \text{ A}$$

The final current is smaller and a longer time is required for it to develop.

(b)   The total inductance and total resistance of the combination of the two inductors are respectively

$$L = 0.2 \text{ H} + 0.2 \text{ H} = 0.4 \text{ H} \qquad R = 3 \ \Omega + 3 \ \Omega = 6 \ \Omega$$

The time constant is now

$$T = \frac{L}{R} = \frac{0.4 \text{ H}}{6 \ \Omega} = 0.067 \text{ s}$$

which is the same as for one of the inductors by itself when short-circuited, as in Problem 9.11.

Supplementary Problems

9.13. Find the inductance in air of a solenoid 2 cm long and 6 mm in diameter that has 500 turns of wire.

9.14. Find the inductance in air of a solenoid 40 cm long and 4 cm in diameter that has 1000 turns of wire.

9.15. A 1-mH inductor is to be made by winding wire on a tube 2 cm in diameter and 10 cm long. How many turns are needed?

9.16. Find the emf induced in a 0.4-mH coil when the current in it is changing at the rate of 500 A/s.

9.17. What rate of change of current is needed to induce an emf of 8 V in a 0.1-H coil?

9.18. An average emf of 32 V is induced in a circuit in which the current drops from 10 A to 2 A in 0.1 s. What is the inductance of the circuit?

9.19. A 2-H coil carries a current of 0.5 A. (*a*) How much energy is stored in it? (*b*) In how much time should the current drop to 0 if an emf of 100 V is to be induced in it?

9.20. What should the current in a 50-mH coil be in order that it contain 0.004 J of energy?

9.21. A potential difference of 100 V is applied to a 50-mH, 40-Ω inductor. (*a*) What is the initial rate at which the current increases? (*b*) What is the rate at which the current is increasing when the current is 1 A? (*c*) What is the final current?

9.22. A potential difference of 50 V is applied across a 12-mH, 8-Ω inductor. (*a*) What is the initial rate at which the current increases? (*b*) What is the current when the rate of change of current is 2000 A/s? (*c*) What is the final current?

9.23. What is the time constant of a 50-mH, 3-Ω inductor?

9.24. A 60-mH, 5-Ω inductor is connected to a 12-V battery whose internal resistance is 1 Ω. (*a*) Find the time constant of the circuit. (*b*) Find the final current in the circuit. (*c*) Find the current 0.005 s, 0.01 s, and 0.05 s after the connection is made.

9.25. Find the current in the above inductor 0.005 s, 0.01 s, and 0.05 s after it has been short-circuited after having been connected to the battery for a long time.

9.26. A 0.1-H, 4-Ω inductor is connected in series with a 0.2 H, 6-Ω inductor and the combination is placed across a 24-V battery whose internal resistance is 2 Ω. (*a*) Find the time constant of the circuit. (*b*) Find the final current in the circuit. (*c*) Find the current 0.01 s and 0.1 s after the connection is made.

9.27. Find the current in the inductors of Problem 9.26 0.01 s and 0.1 s after the battery has been short-circuited after the inductors have been connected to it for a long time.

Answers to Supplementary Problems

9.13. 0.445 mH

9.14. 3.96 mH

9.15. 503 turns

9.16. 0.2 V

9.17. 80 A/s

9.18. 0.4 H

9.19. (*a*) 0.25 J   (*b*) 0.01 s

9.20. 0.4 A

9.21. (*a*) 2000 A/s   (*b*) 1200 A/s   (*c*) 2.5 A

9.22. (*a*) 4167 A/s   (*b*) 3.25 A   (*c*) 6.25 A

9.23. 0.0167 s

9.24. (*a*) 0.01 s   (*b*) 2 A   (*c*) 0.787 A; 1.264 A; 1.987 A

9.25. 1.318 A; 0.869 A; 0.031 A

9.26. (*a*) 0.025 s   (*b*) 2 A   (*c*) 0.659 A; 1.963 A

9.27. 1.433 A; 0.071 A

Chapter 10

Capacitance

CAPACITANCE

A *capacitor* is a system that stores energy in the form of an electric field. In its simplest form a capacitor consists of a pair of parallel metal plates separated by air or other insulating material.

The potential difference V between the plates of a capacitor is directly proportional to the charge Q on either of them, so the ratio Q/V is always the same for a particular capacitor. This ratio is called the *capacitance C* of the capacitor:

$$C = \frac{Q}{V}$$

$$\text{Capacitance} = \frac{\text{charge on either plate}}{\text{potential difference between plates}}$$

The unit of capacitance is the *farad* (F), where 1 farad = 1 coulomb/volt. Since the farad is too large for practical purposes, the *microfarad* and *picofarad* are commonly used, where

$$1 \text{ microfarad} = 1\mu\text{F} = 10^{-6}\text{ F} \qquad 1 \text{ picofarad} = 1\text{pF} = 10^{-12}\text{ F}$$

A charge of 10^{-6} C on each plate of a 1-μF capacitor will produce a potential difference of $V = Q/C = 1$ V between the plates.

A capacitor that consists of parallel plates each of area A separated by the distance d has a capacitance of

$$C = K\epsilon_0 \frac{A}{d}$$

The constant ϵ_0 is the permittivity of free space and its value is

$$\epsilon_0 = 8.85 \times 10^{-12}\text{ F/m}$$

The quantity K is the *dielectric constant* of the material between the capacitor plates; the greater K is, the more effective the material is in diminishing an electric field. For free space, $K = 1$; for air, $K = 1.0006$; a typical value for glass is $K = 6$; and for water, $K = 80$.

Problem 10.1. A 200-pF capacitor is connected to a 100-V battery. Find the charge on the capacitor's plates.

$$Q = CV = 200 \times 10^{-12}\text{ F} \times 100\text{ V} = 2 \times 10^{-8}\text{ C}$$

Problem 10.2. A capacitor has a charge of 5×10^{-4} C when the potential difference across its plates is 300 V. Find its capacitance.

$$C = \frac{Q}{V} = \frac{5 \times 10^{-4}\text{ C}}{300\text{ V}} = 1.67 \times 10^{-6}\text{ F} = 1.67\ \mu\text{F}$$

Problem 10.3. A parallel-plate capacitor has plates 5 cm square and 0.1 mm apart. Find its capacitance (*a*) in air and (*b*) with mica of $K = 6$ between the plates.

(*a*) In air K is very nearly 1, and so

$$C = K\epsilon_0 \frac{A}{d} = 1 \times 8.85 \times 10^{-12}\ \frac{\text{F}}{\text{m}} \times \frac{(0.05\text{ m})^2}{10^{-4}\text{ m}} = 2.21 \times 10^{-10}\text{ F} = 221\text{ pF}$$

(b) With mica between the plates the capacitance will be $K = 6$ times greater, or

$$C = 6 \times 221 \text{ pF} = 1326 \text{ pF}$$

Problem 10.4.   A parallel-plate capacitor has a capacitance of 2 μF in air and 4.6 μF when immersed in benzene.   What is the dielectric constant of benzene?

Since C is proportional to K, in general

$$\frac{C_1}{K_1} = \frac{C_2}{K_2}$$

for the same capacitor.   Here, with $K_1 = K_{air} = 1$,   the dielectric constant K_2 of benzene is

$$K_2 = K_1 \frac{C_2}{C_1} = 1 \times \frac{4.6 \ \mu F}{2 \ \mu F} = 2.3$$

Problem 10.5.   A 10-μF capacitor with air between its plates is connected to a 50-V source and then disconnected.   (a) What are the charge on the capacitor and the potential difference across it?   (b) The space between the plates of the charged capacitor is filled with Teflon   ($K = 2.1$).   What are the charge on the capacitor and the potential difference across it now?

(a)   The capacitor's charge is

$$Q = CV = 10 \times 10^{-6} \text{ F} \times 50 \text{ V} = 5 \times 10^{-4} \text{ C}$$

The potential difference across it remains 50 V after it is disconnected.

(b)   The presence of another dielectric does not change the charge on the capacitor.   Since its capacitance is now

$$C_2 = \frac{K_2}{K_1} C_1$$

and $V = Q/C$,   the new potential difference is

$$V_2 = \frac{Q}{C_2} = \frac{K_1}{K_2} \frac{Q}{C_1} = \frac{K_1}{K_2} V_1 = \frac{1}{2.1} \times 50 \text{ V} = 23.8 \text{ V}$$

ENERGY OF A CHARGED CAPACITOR

In order to produce the electric field in a charged capacitor, work must be done to separate the positive and negative charges.   This work is stored as electric potential energy in the capacitor.   The potential energy W of a capacitor of capacitance C whose charge is Q and whose potential difference is V is given by

$$W = \frac{1}{2} QV = \frac{1}{2} CV^2 = \frac{1}{2} \frac{Q^2}{C}$$

Problem 10.6.   How much energy is stored in a 50-pF capacitor when it is charged to a potential difference of 200 V?

$$W = \frac{1}{2} CV^2 = \frac{1}{2} \times 50 \times 10^{-12} \text{ F} \times (200 \text{ V})^2 = 10^{-6} \text{ J}$$

Problem 10.7.   A 100-μF capacitor is to have an energy content of 50 J in order to operate a flashlamp.   (a) What voltage is required to charge the capacitor?   (b) How much charge passes through the flashlamp?

(a)   Since $W = (1/2)CV^2$,

$$V = \sqrt{\frac{2W}{C}} = \sqrt{\frac{2 \times 50 \text{ J}}{10^{-4} \text{ F}}} = 1000 \text{ V}$$

(b)

$$Q = CV = 10^{-4}\,\text{F} \times 10^3\,\text{V} = 0.1\,\text{C}$$

CAPACITORS IN SERIES

The *equivalent capacitance* of a set of capacitors connected together is the capacitance of the single capacitor that can replace the set without changing the properties of any circuit. Figure 10-1 shows three capacitors connected in series. Each one has charges of the same magnitude Q on its plates because of the way they are joined, and the voltages across them are

$$V_1 = \frac{Q}{C_1} \qquad V_2 = \frac{Q}{C_2} \qquad V_3 = \frac{Q}{C_3}$$

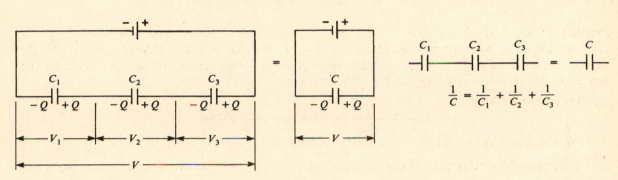

Fig. 10-1

If C is the equivalent capacitance of the set, then

$$V = V_1 + V_2 + V_3 \qquad \frac{Q}{C} = \frac{Q}{C_1} + \frac{Q}{C_2} + \frac{Q}{C_3}$$

Dividing through by the charge Q gives

$$\frac{1}{C} = \frac{1}{C_1} + \frac{1}{C_2} + \frac{1}{C_3}$$

For any number of capacitors in series, the same rule holds: $1/C$ is equal to the sum of the $1/C$ values for the set. In symbols,

$$\frac{1}{C} = \frac{1}{C_1} + \frac{1}{C_2} + \frac{1}{C_3} + \cdots$$

If there are only two capacitors in series,

$$\frac{1}{C} = \frac{1}{C_1} + \frac{1}{C_2} = \frac{C_1 + C_2}{C_1 C_2}$$

and so

$$C = \frac{C_1 C_2}{C_1 + C_2}$$

Problem 10.8. Three capacitors whose capacitances are 1 μF, 2 μF, and 3 μF are connected in series. Find the equivalent capacitance of the combination.

$$\frac{1}{C} = \frac{1}{C_1} + \frac{1}{C_2} + \frac{1}{C_3} = \frac{1}{1\,\mu\text{F}} + \frac{1}{2\,\mu\text{F}} + \frac{1}{3\,\mu\text{F}} = \frac{11}{6\,\mu\text{F}}$$

Hence

$$11C = 6 \ \mu F$$

$$C = \frac{6}{11} \ \mu F = 0.545 \ \mu F$$

Problem 10.9. A 2-μF and a 3-μF capacitor are connected in series. What is their equivalent capacitance?

$$C = \frac{C_1 C_2}{C_1 + C_2} = \frac{2 \ \mu F \times 3 \ \mu F}{2 \ \mu F + 3 \ \mu F} = 1.2 \ \mu F$$

Problem 10.10. A potential difference of 500 V is applied across the capacitors of Problem 10.9, as in Fig. 10-2. Find the charge on each capacitor and the potential difference across it.

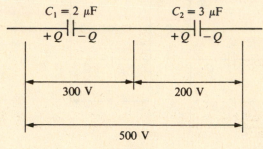

Fig. 10-2

The charge on the combination is

$$Q = CV = 1.2 \times 10^{-6} \ F \times 500 \ V = 6 \times 10^{-4} \ C$$

The same charge is present on each capacitor. The potential difference across the 2-μF capacitor is

$$V_1 = \frac{Q}{C_1} = \frac{6 \times 10^{-4} \ C}{2 \times 10^{-6} \ F} = 300 \ V$$

and that across the 3-μF capacitor is

$$V_2 = \frac{Q}{C_2} = \frac{6 \times 10^{-4} \ C}{3 \times 10^{-6} \ F} = 200 \ V$$

As a check we note that $V_1 + V_2 = 500$ V.

CAPACITORS IN PARALLEL

Figure 10-3 shows three capacitors connected in parallel. Now the same voltage V is across all of them, and their respective charges are

$$Q_1 = C_1 V \qquad Q_2 = C_2 V \qquad Q_3 = C_3 V$$

The total charge $Q_1 + Q_2 + Q_3$ on either the + or − plates of the capacitors is equal to the charge Q on the corresponding plate of the equivalent capacitor C, and so

$$Q = Q_1 + Q_2 + Q_3 \qquad CV = C_1 V + C_2 V + C_3 V$$

Dividing through by V gives

$$C = C_1 + C_2 + C_3$$

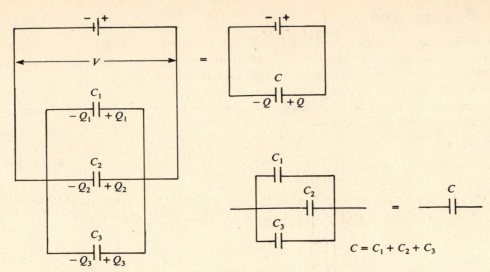

Fig. 10-3

For more than three capacitors,

$$C = C_1 + C_2 + C_3 + \cdots$$

Problem 10.11.   The three capacitors of Problem 10.8 are connected in parallel.   Find the equivalent capacitance of the combination.

$$C = C_1 + C_2 + C_3 = 1 \ \mu\text{F} + 2 \ \mu\text{F} + 3 \ \mu\text{F} = 6 \ \mu\text{F}$$

Problem 10.12.   A 5-pF and a 10-pF capacitor are connected in parallel.   (a) What is their equivalent capacitance?   (b) A potential difference of 1000 V is applied to the combination.   Find the charge on each capacitor and the potential difference across it.

(a) $C = C_1 + C_2 = 5 \ \text{pF} + 10 \ \text{pF} = 15 \ \text{pF}$

(b)   The same potential difference $V = 1000 \ \text{V}$ is across each capacitor (Fig. 10-4).   The charge on the 5-pF capacitor is

$$Q_1 = C_1 V = 5 \times 10^{-12} \ \text{F} \times 10^3 \ \text{V} = 5 \times 10^{-9} \ \text{C}$$

and that on the 10-pF capacitor is

$$Q_2 = C_2 V = 10 \times 10^{-12} \ \text{F} \times 10^3 \ \text{V} = 10^{-8} \ \text{C}$$

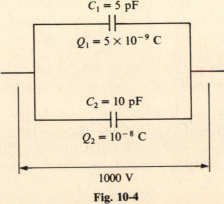

Fig. 10-4

CHARGING A CAPACITOR

When a capacitor is being charged in a circuit such as that of Fig. 10-5, at any moment the voltage Q/C across it is in the opposite direction to the battery voltage V and thus tends to oppose the flow of additional charge. For this reason a capacitor does not acquire its final charge the instant it is connected to a battery or other source of emf. The net potential difference when the charge on the capacitor is Q is $V - Q/C$ and the current is then

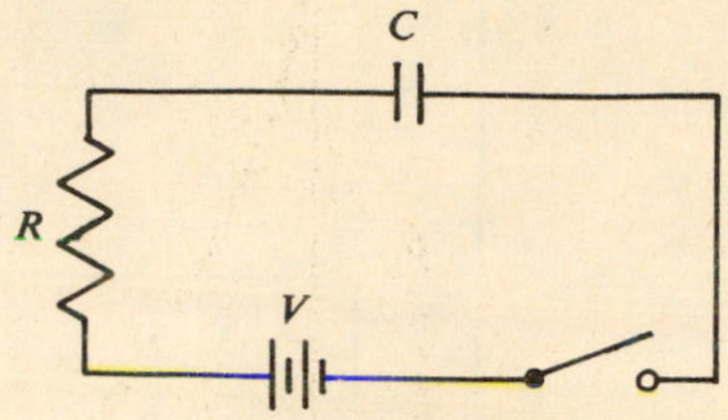

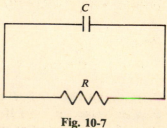

Fig. 10-5

$$I = \frac{\Delta Q}{\Delta t} = \frac{V - Q/C}{R}$$

As Q increases, its rate of increase $I = \Delta Q/\Delta t$ decreases. Figure 10-6 shows how Q varies with time when a capacitor is being charged; the switch of Fig. 10-5 is closed at $t = 0$.

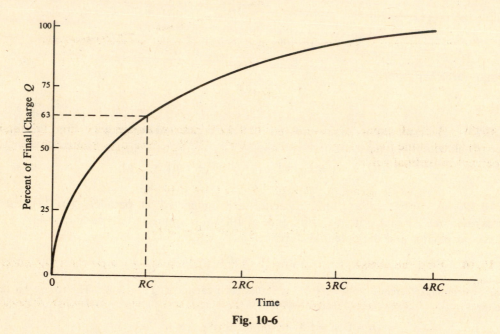

Fig. 10-6

The product RC of the resistance in the circuit and the capacitance C governs the rate at which the capacitor reaches its ultimate charge of $Q_0 = CV$. The product RC is called the *time constant* T of the circuit. After a time equal to T, the charge on the capacitor will be 63 percent of its final value.

If a charged capacitor is discharged through a resistance R, as in Fig. 10-7, its charge will fall to 37 percent of its original value after $T = RC$ (Fig. 10-8). The smaller the time constant T, the more rapidly a capacitor can be charged or discharged. The curves of Figs. 10-6 and 10-8 are the same in shape as those of Figs. 9-3 and 9-4.

The formula that governs the growth of charge in the circuit of Fig. 10-5 is

Fig. 10-7

$$Q = Q_0(1 - e^{-t/T})$$

where Q_0 is the final charge CV and T is the time constant RC. Figure 10-6 is a graph of this formula. We note the resemblance to the formula $I = I_0(1 - e^{-t/T})$ from Chapter 9.

When a charged capacitor is discharged through a resistance, as in Fig. 10-7, the decrease in charge is governed by the formula

$$Q = Q_0 e^{-t/T}$$

where again $T = RC$ is the time constant.

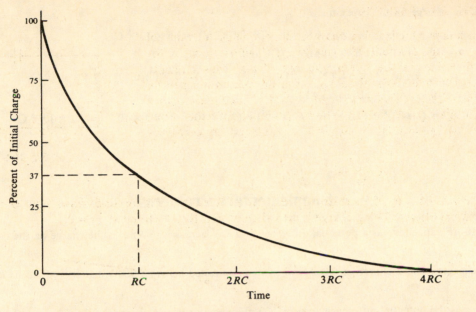

Fig. 10-8

Problem 10.13. A 20-μF capacitor is connected to a 45-V battery through a circuit whose resistance is 2000 Ω. (a) What is the final charge on the capacitor? (b) How long does it take for the charge to reach 63 percent of its final value?

(a) $$Q_0 = CV = 20 \times 10^{-6}\,\text{F} \times 45\,\text{V} = 9 \times 10^{-4}\,\text{C}$$

(b) $$T = RC = 2000\,\Omega \times 20 \times 10^{-6}\,\text{F} = 0.04\,\text{s}$$

Problem 10.14. Find the charge on the above capacitor 0.01 s and 0.1 s after the connection to the battery is made.

When $t = 0.01$ s, $t/T = (0.01\,\text{s})/(0.04\,\text{s}) = 0.25$.   Using a calculator or table of exponentials gives
$$e^{-t/T} = e^{-0.25} = 0.78$$
Hence
$$Q = Q_0(1 - e^{-t/T}) = (9 \times 10^{-4}\,\text{C})(1 - 0.78) = (9 \times 10^{-4}\,\text{C})(0.22) = 2.0 \times 10^{-4}\,\text{C}$$
Similarly, when $t = 0.1$ s, $t/T = (0.1\,\text{s})/(0.04\,\text{s}) = 2.5$ and
$$e^{-t/T} = e^{-2.5} = 0.082$$
Hence
$$Q = Q_0(1 - e^{-t/T}) = (9 \times 10^{-4}\,\text{C})(1 - 0.082) = (9 \times 10^{-4}\,\text{C})(0.918) = 8.3 \times 10^{-4}\,\text{C}$$

Problem 10.15. A 5-μF capacitor is charged by being connected to a 3-V battery. The battery is then disconnected. (a) If the resistance of the dielectric material between the capacitor plates is $10^9\,\Omega$, find the period of time required for the charge on the capacitor to drop to 37 percent of its original value. (b) What is the charge remaining on the capacitor 1 h after it has been disconnected? What is the charge 10 h afterward?

(a) $$T = RC = (10^9\,\Omega)(5 \times 10^{-6}\,\text{F}) = 5 \times 10^3\,\text{s}$$
which is
$$\frac{5 \times 10^3\,\text{s}}{60\,\text{s/min} \times 60\,\text{min/h}} = \frac{5000\,\text{s}}{3600\,\text{s/h}} = 1.4\,\text{h}$$

(b) The initial charge on the capacitor is

$$Q_0 = CV = (5 \times 10^{-6} \text{ F})(3 \text{ V}) = 1.5 \times 10^{-5} \text{ C}$$

After $t = 1$ h, $t/T = (1 \text{ h})/(1.4 \text{ h}) = 0.71$, and

$$Q = Q_0 e^{-t/T} = (1.5 \times 10^{-5} \text{ C})(e^{-0.71}) = (1.5 \times 10^{-5} \text{ C})(0.49) = 7.4 \times 10^{-6} \text{ C}$$

After $t = 10$ h, $t/T = (10 \text{ h})/(1.4 \text{ h}) = 7.1$, and

$$Q = Q_0 e^{-t/T} = (1.5 \times 10^{-5} \text{ C})(e^{-7.1}) = (1.5 \times 10^{-5} \text{ C})(8.3 \times 10^{-4}) = 1.2 \times 10^{-8} \text{ C}$$

Supplementary Problems

10.16. A 10-μF capacitor has a potential difference of 250 V across it. What is the charge on the capacitor?

10.17. A capacitor has a charge of 0.002 C when it is connected across a 100-V battery. Find its capacitance.

10.18. What is the potential difference across a 500-pF capacitor whose charge is 0.3 μC?

10.19. The plates of a parallel-plate capacitor have areas of 40 cm^2 and are separated by 0.2 mm of waxed paper of $K = 2.2$. Find its capacitance.

10.20. The waxed paper is removed from between the plates of the above capacitor. Find the new capacitance.

10.21. The plates of a parallel-plate capacitor of capacitance C are moved closer together until they are half their original separation. What is the new capacitance?

10.22. The capacitance of a parallel-plate capacitor is increased from 8 μF to 50 μF when a sheet of glass is inserted between its plates. What is the dielectric constant K of the glass?

10.23. Find the energy stored in a 5-μF capacitor when it is charged to a potential difference of 1000 V.

10.24. (a) What potential difference must be applied across a 10-μF capacitor if it is to have an energy content of 1 J? (b) What is the charge on the capacitor under these circumstances?

10.25. Three capacitors whose capacitances are 5 μF, 10 μF, and 20 μF are connected in series. Find the equivalent capacitance of the combination.

10.26. The capacitors of Problem 10.25 are connected in parallel. Find the equivalent capacitance of the combination.

10.27. List the capacitances that can be obtained by combining three 10-μF capacitors in all possible ways.

10.28. A 20-pF capacitor and a 25-pF capacitor are connected in parallel and a potential difference of 100 V is applied to the combination. (a) Find the equivalent capacitance. (b) Find the charge on each capacitor. (c) Find the potential difference across each capacitor.

10.29. A 50-pF capacitor and a 75-pF capacitor are connected in series and a potential difference of 250 V is applied to the combination.   (a) Find the equivalent capacitance.   (b) Find the charge on each capacitor.   (c) Find the potential difference across each capacitor.

10.30.   A 5-μF capacitor is connected to a 100-V battery through a circuit whose resistance is 800 Ω.   (a) What is the time constant of this arrangement?   (b) What is the initial current that flows when the battery is connected?   (c) What is the final charge on the capacitor?

10.31.   Find the charge on the capacitor of Problem 10.30 0.001 s, 0.005 s, and 0.01 s after the connection to the battery is made.

10.32.   The resistance of the dielectric between the plates of the capacitor of Problem 10.30 is $10^7\,\Omega$.   If the capacitor is disconnected from the battery, find the charge remaining on it 30 s, 1 min, and 10 min later.

Answers to Supplementary Problems

10.16. 0.0025 C	**10.20.** 177 pF	**10.24.** (a) 447 V (b) 4.47×10^{-3} C
10.17. 20 μF	**10.21.** 2C	**10.25.** 2.86 μF
10.18. 600 V	**10.22.** 6.25	**10.26.** 35 μF
10.19. 389 pF	**10.23.** 2.5 J	**10.27.** 3.33 μF; 6.67 μF; 15 μF; 30 μF

10.28.   (a) 45 pF   (b) 2×10^{-9} C; 2.5×10^{-9} C   (c) 100 V; 100 V

10.29.   (a) 30 pF   (b) 7.5×10^{-9} C; 7.5×10^{-9} C   (c) 150 V; 100 V

10.30.   (a) 0.004 s   (b) 0.125 A   (c) 5×10^{-4} C

10.31. 1.106×10^{-4} C; 3.567×10^{-4} C; 4.590×10^{-4} C

10.32. 2.744×10^{-4} C; 1.506×10^{-4} C; 3.072×10^{-9} C

Chapter 11

Trigonometry and Vectors

MEASURING ANGLES

In everyday life, angles are measured in *degrees*, where 360° equals a full rotation. A *right angle* contains 90° and represents a quarter of a full rotation. Figure 11-1 illustrates some angles.

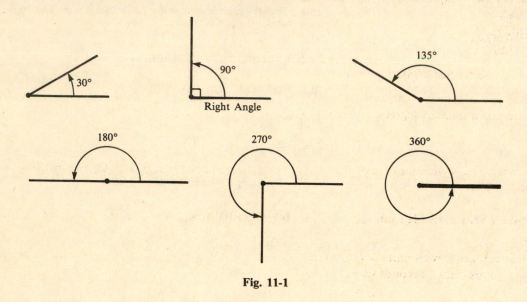

Fig. 11-1

Another angular unit widely used for technical purposes is the *radian* (rad). If a circle is drawn whose center is at the corner of a particular angle, as in Fig. 11-2, the angle in radians is equal to the ratio between the arc s cut by the angle and the radius r of the circle. Using the Greek letter θ (theta) for the angle,

$$\theta = \frac{s}{r}$$

$$\text{Angle in radians} = \frac{\text{arc length}}{\text{radius}}$$

Because the circumference of a circle of radius r is $2\pi r$, there are 2π rad in a complete rotation. Hence $360° = 2\pi$ rad and

$$180° = \pi \text{ rad}$$

This means that we can convert from degree to radian measure or vice versa by using the formulas

$$\theta\ (\text{rad}) = \theta\ (°) \times \frac{\pi \text{ rad}}{180°} = \theta\ (°) \times 0.01745 \text{ rad}/°$$

$$\theta\ (°) = \theta\ (\text{rad}) \times \frac{180°}{\pi \text{ rad}} = \theta\ (\text{rad}) \times 57.30°/\text{rad}$$

An angle of 1 rad is the same as an angle of 57.30°.

104

Problem 11.1.   Express 8° in radians.
$$\theta = 8° \times 0.01745 \text{ rad}/° = 0.14 \text{ rad}$$

Problem 11.2.   Express 2.5 rad in degrees.
$$\theta = 2.5 \text{ rad} \times 57.3°/\text{rad} = 143°$$

Problem 11.3.   Express 90° in radians in terms of π.
$$\theta = 90° \times \frac{\pi \text{ rad}}{180°} = \frac{\pi}{2} \text{ rad}$$

Problem 11.4.   Express $3\pi/2$ rad in degrees.
$$\theta = \frac{3}{2}\pi \text{ rad} \times \frac{180°}{\pi \text{ rad}} = 270°$$

MINUTES AND SECONDS

Fractions of a degree can be expressed in decimal form, so that half a degree would be written 0.5° and four and a quarter degrees would be written 4.25°:
$$\frac{1}{2}° = 0.5° \qquad 4\frac{1}{4}° = 4.25°$$

Another method is commonly used.   In this method a degree is divided into 60 parts called *minutes* (1 minute = 1′), so that
$$60' = 1°$$
A minute is further divided into 60 parts called *seconds* (1 second = 1″), so that
$$60'' = 1'$$
To avoid confusion with units of time, minutes and seconds in angular measure are sometimes called "minutes of arc" and "seconds of arc."

Problem 11.5.   Express an angle of 230°16′45″ in decimal form.

We proceed as follows:
$$45'' = \left(\frac{45}{60}\right)' = 0.75'$$
$$16'45'' = 16' + 45'' = 16' + 0.75' = 16.75'$$
$$16.75' = \left(\frac{16.75}{60}\right)° = 0.28°$$
$$230°16'45'' = 230° + 0.28° = 230.28°$$

Problem 11.6.   Express an angle of 38.17° in degrees, minutes, and seconds.
$$0.17° = (0.17 \times 60)' = 10.2' = 10' + 0.2'$$
Since $0.2' = (0.2 \times 60)'' = 12''$,   we have
$$0.17° = 10'12''$$
and so
$$38.17° = 38°10'12''$$

Problem 11.7.   Find the sum of 121°48′30″ and 6°51′42″.

$$
\begin{array}{r}
121°48'30'' \\
+ \quad 6°51'42'' \\
\hline
127°99'72''
\end{array}
$$

Since $60'' = 1'$ and $60' = 1°$,

$$127°99'72'' = 127°100'12'' = 128°40'12''$$

Problem 11.8.   Subtract $8°52'6''$ from $44°14'3''$.

In order to subtract in this case, we must first make some conversions.   We note that $44° = 43°60'$,   so that we can write

$$44°14'3'' = 43°74'3''$$

Similarly $74' = 73'60''$,   and so

$$44°14'3'' = 43°74'3'' = 43°73'63''$$

Now we can carry out the subtraction in the usual way:

$$
\begin{array}{r}
43°73'63'' \\
-\ 8°52'16'' \\
\hline
35°21'47'' \\
\end{array}
$$

Problem 11.9.   Multiply $8°35'20''$ by 5.

The procedure is to multiply the degrees, minutes, and seconds separately and then to add the products together.

$$8° \times 5 = 40°$$
$$35' \times 5 = 175' = 60' + 60' + 55' = 2°55'$$
$$20'' \times 5 = 100'' = 60'' + 40'' = 1'40''$$

Hence

$$8°35'20'' \times 5 = 40° + 2°55' + 1'40'' = 42°56'40''$$

TRIGONOMETRIC FUNCTIONS

A *right triangle* is a triangle two of whose sides are perpendicular.   The *hypotenuse* of a right triangle is the side opposite the right angle, as in Fig. 11-3; the hypotenuse is always the longest side. The three basic trigonometric functions, the sine, cosine, and tangent of an angle, are defined in terms of the right triangle of Fig. 11-3 as follows:

$$\sin \theta = \frac{a}{c} = \frac{\text{opposite side}}{\text{hypotenuse}}$$

$$\cos \theta = \frac{b}{c} = \frac{\text{adjacent side}}{\text{hypotenuse}}$$

$$\tan \theta = \frac{a}{b} = \frac{\text{opposite side}}{\text{adjacent side}} = \frac{\sin \theta}{\cos \theta}$$

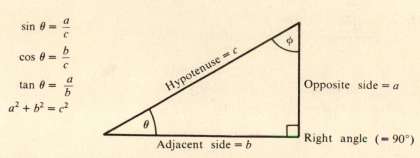

$$\sin \theta = \frac{a}{c}$$
$$\cos \theta = \frac{b}{c}$$
$$\tan \theta = \frac{a}{b}$$
$$a^2 + b^2 = c^2$$

Fig. 11-3

Problem 11.10. Find the values of the sine, cosine, and tangent of angle θ in Fig. 11-4.

$$\sin \theta = \frac{\text{opposite side}}{\text{hypotenuse}} = \frac{3 \text{ cm}}{5 \text{ cm}} = 0.6$$

$$\cos \theta = \frac{\text{adjacent side}}{\text{hypotenuse}} = \frac{4 \text{ cm}}{5 \text{ cm}} = 0.8$$

$$\tan \theta = \frac{\text{opposite side}}{\text{adjacent side}} = \frac{3 \text{ cm}}{4 \text{ cm}} = 0.75$$

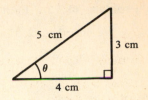

Fig. 11-4

The *inverse* of a trigonometric function is the angle whose function is given. Thus the inverse of $\sin \theta$ is the angle θ. The names and abbreviations of the inverse trigonometric functions are as follows:

$$\sin \theta = x$$
$$\theta = \arcsin x = \sin^{-1} x = \text{angle whose sine is } x$$
$$\cos \theta = y$$
$$\theta = \arccos y = \cos^{-1} y = \text{angle whose cosine is } y$$
$$\tan \theta = z$$
$$\theta = \arctan z = \tan^{-1} z = \text{angle whose tangent is } z$$

It is important to remember that in trigonometry an expression such as $\sin^{-1} x$ does *not* signify $1/\sin x$, even though in algebra the exponent -1 signifies a reciprocal.

Problem 11.11. Find the angle whose cosine is 0.952.

With some calculators, the procedure is to enter 0.952 and then press the key marked $\cos^{-1}$ (or arccos). The result will appear as 17.8242, which can be rounded off to 18°. In a table of cosines, such as the one in Appendix E, we look for the value nearest to 0.952 and then read across to find the corresponding angle. What we would find is that $\cos 17° = 0.956$ and $\cos 18° = 0.951$. Since 0.951 is closer to 0.952 than 0.956 is, to the nearest degree we have $\cos^{-1} 0.952 = 18°$.

LARGE ANGLES

Numerical tables of the sine, cosine, and tangent of angles from 0° to 90° are given Appendix E. It is possible to extend the definitions of these functions to cover angles from 90° to 360°, and the tables of Appendix E may be used for such angles with the help of the relationships in Table 11-1.

Table 11-1

	θ	$90° + \theta$	$180° + \theta$	$270° + \theta$
sine	$\sin \theta$	$\cos \theta$	$-\sin \theta$	$-\cos \theta$
cosine	$\cos \theta$	$-\sin \theta$	$-\cos \theta$	$\sin \theta$
tangent	$\tan \theta$	$-\dfrac{1}{\tan \theta}$	$\tan \theta$	$-\dfrac{1}{\tan \theta}$

Problem 11.12. Find the value of sin 120°.

We begin by noting that $120° = 90° + 30°$. Then, since from Table 11-1

$$\sin(90° + \theta) = \cos \theta$$

we have

$$\sin 120° = \sin(90° + 30°) = \cos 30° = 0.866$$

Problem 11.13. Find the value of $\cos 250°$.

We note that $250° = 180° + 70°$. From Table 11-1

$$\cos(180° + \theta) = -\cos\theta$$

and so

$$\cos 250° = \cos(180° + 70°) = -\cos 70° = -0.342$$

Problem 11.14. Find the value of $\tan 342°$.

We note that $342° = 270° + 72°$. From Table 11-1

$$\tan(270° + \theta) = -\frac{1}{\tan\theta}$$

and so

$$\tan 342° = \tan(270° + 72°) = -\frac{1}{\tan 72°} = -\frac{1}{3.078} = -0.325$$

PYTHAGOREAN THEOREM

The Pythagorean theorem is a useful relationship that holds in a right triangle. This theorem states that the sum of the squares of the short sides of a right triangle is equal to the square of the hypotenuse (longest side). For the triangle of Fig. 11-3,

$$a^2 + b^2 = c^2$$

Hence we can always express the length of any of the sides of a right triangle in terms of the other sides:

$$a = \sqrt{c^2 - b^2} \qquad b = \sqrt{c^2 - a^2} \qquad c = \sqrt{a^2 + b^2}$$

Problem 11.15. Find the length of the hypotenuse of a right triangle whose short sides are 6 in and 12 in long.

$$c = \sqrt{a^2 + b^2} = \sqrt{6^2 + 12^2}\text{ in} = \sqrt{36 + 144}\text{ in} = \sqrt{180}\text{ in} = 13.4\text{ in}$$

Problem 11.16. The hypotenuse of a right triangle is 8 m long and one of the short sides is 6 m long. Find the length of the other short side.

$$a = \sqrt{c^2 - b^2} = \sqrt{8^2 - 6^2}\text{ m} = \sqrt{64 - 36}\text{ m} = \sqrt{28}\text{ m} = 5.3\text{ m}$$

SOLVING A RIGHT TRIANGLE

To solve a given triangle means to find the values of any unknown sides or angles in terms of the values of the known sides and angles. Of the six quantities that characterize a triangle—three sides and three angles—we must know the values of at least three, including one of the sides, in order to solve the triangle for the others. In a right triangle, one of the angles is 90° by definition, so all we need are the lengths of any two of its sides or the length of one side plus the value of one of the other angles to find the other sides and angles. In this book we need consider only right triangles.

Suppose we know the length of side b and the value of angle θ in the right triangle of Fig. 11-5. From the definitions of sine and tangent, we see that

$$\tan\theta = \frac{a}{b} \qquad \sin\theta = \frac{a}{c}$$

$$a = b\tan\theta \qquad c = \frac{a}{\sin\theta}$$

This gives us the two unknown sides. To find the unknown angle

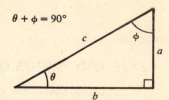

$\theta + \phi = 90°$

Fig. 11-5

ϕ (Greek letter "phi"), we can use any one of these formulas:

$$\phi = \sin^{-1}\frac{b}{c} \qquad \phi = \cos^{-1}\frac{a}{c} \qquad \phi = \tan^{-1}\frac{b}{a}$$

A useful relationship in trigonometry is that the sum of the angles of any triangle is 180°. Since one of the angles in a right triangle is 90°, the sum of the other two must be 90°:

$$\theta + \phi = 90°$$

Problem 11.17. Find the sides a and b and the angle ϕ in the triangle of Fig. 11-5 when $c = 8$ in and $\theta = 30°$.

We start with the length of side a:

$$\sin\theta = \frac{a}{c}$$

$$\sin 30° = \frac{a}{8\text{ in}}$$

$$a = 8\text{ in} \times \sin 30° = 8\text{ in} \times 0.500 = 4\text{ in}$$

To find b we can proceed in two ways. One is this:

$$\cos\theta = \frac{b}{c}$$

$$\cos 30° = \frac{b}{8\text{ in}}$$

$$b = 8\text{ in} \times \cos 30° = 8\text{ in} \times 0.866 = 6.93\text{ in}$$

The other way is to use the Pythagorean theorem:

$$b = \sqrt{c^2 - a^2} = \sqrt{8^2 - 4^2}\text{ in} = \sqrt{64 - 16}\text{ in} = \sqrt{48}\text{ in} = 6.93\text{ in}$$

Since $\theta + \phi = 90°$ in a right triangle, here $\phi = 60°$ since we are given that $\theta = 30°$.

Problem 11.18. In the triangle of Fig. 11-5, $a = 7$ m and $b = 10$ m. Find the value of c, θ, and ϕ.

To find c, we use the Pythagorean theorem:

$$c = \sqrt{a^2 + b^2} = \sqrt{7^2 + 10^2}\text{ m} = \sqrt{49 + 100}\text{ m} = \sqrt{149}\text{ m} = 12.2\text{ m}$$

To find θ, we proceed in this way:

$$\tan\theta = \frac{a}{b} = \frac{7\text{ m}}{10\text{ m}} = 0.7$$

$$\theta = \tan^{-1} 0.7 = 35°$$

Since $\theta = 35°$,

$$\phi = 90° - 35° = 55°$$

Problem 11.19. In the triangle of Fig. 11-5, $a = 3$ cm and $c = 9$ cm. Find the values of b, θ, and ϕ.

$$b = \sqrt{c^2 - a^2} = \sqrt{9^2 - 3^2}\text{ cm} = \sqrt{81 - 9}\text{ cm} = \sqrt{72}\text{ cm} = 8.5\text{ cm}$$

$$\sin\theta = \frac{a}{c} = \frac{3\text{ cm}}{9\text{ cm}} = 0.333$$

$$\theta = \sin^{-1} 0.333 = 19°$$

$$\phi = 90° - \theta = 90° - 19° = 71°$$

SCALAR AND VECTOR QUANTITIES

A *scalar quantity* has only magnitude and is completely specified by a number and a unit. Examples are mass (a stone has a mass of 2 kg), volume (a bottle has a volume of 12 oz), and frequency (house current has a frequency of 60 cycles/s). Symbols of scalar quantities are printed in italic type ($m =$ mass, $V =$ volume).   Scalar quantities of the same kind are added using ordinary arithmetic.

A *vector quantity* has both magnitude and direction.   Examples are displacement (an airplane has flown 200 mi to the southwest), velocity (a car is moving at 60 mi/h to the north), and force (a man applies an upward force of 15 lb to a package).   Symbols of vector quantities are printed in boldface type ($\mathbf{v} =$ velocity, $\mathbf{F} =$ force) and expressed in handwriting by arrows over the letters ($\vec{v}, \vec{F}$).   The magnitude of a vector quantity is printed in italic type (F is the magnitude of the force $\mathbf{F}$).   When vector quantities are added, their directions must be taken into account.

VECTOR ADDITION

A *vector* is an arrowed line whose length is proportional to a certain vector quantity and whose direction indicates the direction of the quantity.

To add the vector $\mathbf{B}$ to the vector $\mathbf{A}$, draw $\mathbf{B}$ so that its tail is at the head of $\mathbf{A}$.   The vector sum $\mathbf{A} + \mathbf{B}$ is the vector $\mathbf{R}$ that joins the tail of $\mathbf{A}$ and the head of $\mathbf{B}$ (Fig. 11-6). $\mathbf{R}$ is usually called the *resultant* of $\mathbf{A}$ and $\mathbf{B}$.

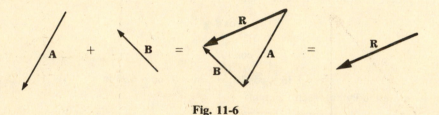

Fig. 11-6

The order in which $\mathbf{A}$ and $\mathbf{B}$ are added is not significant, so that $\mathbf{A} + \mathbf{B} = \mathbf{B} + \mathbf{A}$ (Figs. 11-6 and 11-7).

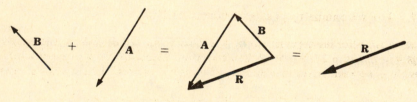

Fig. 11-7

Problem 11.20.   A man walks eastward for 5 miles and then northward for 10 miles.   How far is he from his starting point?   If he had walked directly to his destination, in what direction would he have headed?

From Fig. 11-8, the length of the resultant vector $\mathbf{R}$ corresponds to a distance of 11.2 miles and a protractor shows that its direction is 27° east of north.

It is easy to apply trigonometry to find the resultant $\mathbf{R}$ of two vectors $\mathbf{A}$ and $\mathbf{B}$ that are perpendicular to each other.   The magnitude of the resultant is given by the Pythagorean theorem as

$$R = \sqrt{A^2 + B^2}$$

and the angle θ between $\mathbf{R}$ and $\mathbf{A}$ (Fig. 11-9) may be found from the relationship

$$\tan \theta = \frac{B}{A}$$

by examining a table of tangents to find the angle whose tangent is closest in value to B/A.

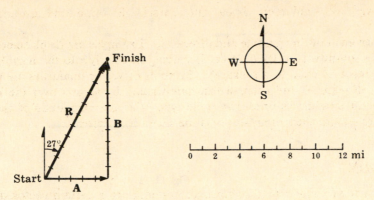

Fig. 11-8

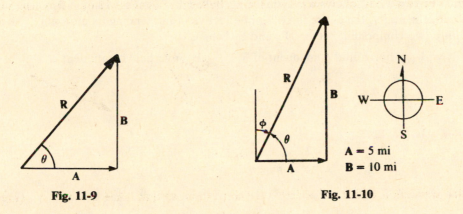

Fig. 11-9   Fig. 11-10

Problem 11.21.   Use trigonometry to solve Problem 11.20.

From a trigonometric table (for instance the one in Appendix E) or with the help of a calculator we find that the angle whose tangent is closest to 2 is $\theta = 63°$.   To express the direction of **R** in terms of north, we see from Fig. 11-10 that the angle ϕ between north and **R**, plus the angle θ between **R** and east, is equal to 90°.   Since $\phi + \theta = 90°$,

$$R = \sqrt{A^2 + B^2} = \sqrt{(5 \text{ mi})^2 + (10 \text{ mi})^2} = \sqrt{25 \text{ mi}^2 + 100 \text{ mi}^2} = \sqrt{125 \text{ mi}^2} = 11.2 \text{ mi}$$

To find the direction of **R** we note that

$$\tan \theta = \frac{B}{A} = \frac{10 \text{ mi}}{5 \text{ mi}} = 2$$

From a trigonometric table (for instance the one in Appendix E) or with the help of a calculator we find that the angle whose tangent is closest to 2 is $\theta = 63°$.   To express the direction of **R** in terms of north, we see from Fig. 11-10 that the angle ϕ between north and **R**, plus the angle θ between **R** and east, is equal to 90°.   Since $\phi + \theta = 90°$,

$$\phi = 90° - \theta = 90° - 63° = 27°$$

The resultant **R** has a magnitude of 11.2 mi and its direction is 27° east of north.   This is the same as the answer obtained graphically in Problem 11.20.

RESOLVING A VECTOR

Just as two or more vectors can be added together to yield a single resultant vector, so it is possible to break up a single vector into two or more other vectors.   If the vectors **A** and **B** are together equivalent to the vector **C**, then the vector **C** is equivalent to the two vectors **A** and **B** (Fig. 11-11). When a vector is replaced by two or more others, the process is called *resolving* the vector, and the new vectors are known as the *components* of the initial vector.

Vector
addition

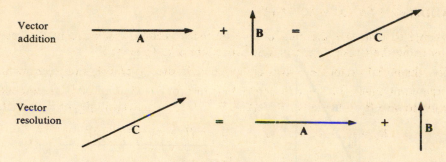

Vector
resolution

Fig. 11-11

The components into which a vector is resolved are nearly always chosen to be perpendicular to one another. Figure 11-12 shows a wagon being pulled by a man with the force **F**. Because the wagon moves horizontally, the entire force is not effective in influencing its motion. The force **F** may be resolved into two component vectors, $\mathbf{F}_x$ and $\mathbf{F}_y$, where

$\mathbf{F}_x$ = horizontal component of **F** $\mathbf{F}_y$ = vertical component of **F**

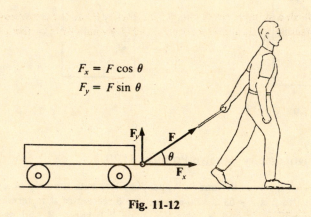

$F_x = F \cos \theta$
$F_y = F \sin \theta$

Fig. 11-12

The magnitudes of these components are

$$F_x = F \cos \theta \qquad F_y = F \sin \theta$$

Evidently the component $\mathbf{F}_x$ is responsible for the wagon's motion, and if we were interested in working out the details of this motion, we would need to consider only $\mathbf{F}_x$.

Problem 11.22. The man in Fig. 11-12 exerts a force of 10.0 lb on the wagon at an angle of $\theta = 30°$ above the horizontal. Find the horizontal and vertical components of this force.

The magnitudes of $\mathbf{F}_x$ and $\mathbf{F}_y$ are respectively

$$F_x = F \cos \theta = 10.0 \text{ lb} \times \cos 30° = 8.66 \text{ lb} \qquad F_y = F \sin \theta = 10.0 \text{ lb} \times \sin 30° = 5.00 \text{ lb}$$

We note that $F_x + F_y = 13.66$ lb although **F** itself has the magnitude $F = 10.0$ lb. What is wrong? The answer is that nothing is wrong; because F_x and F_y are just the *magnitudes* of the vectors $\mathbf{F}_x$ and $\mathbf{F}_y$, it is meaningless to add them together. However, we can certainly add the *vectors* $\mathbf{F}_x$ and $\mathbf{F}_y$ to find the magnitude of their resultant **F**. Because $\mathbf{F}_x$ and $\mathbf{F}_y$ are perpendicular,

$$F = \sqrt{F_x^2 + F_y^2} = \sqrt{(8.66 \text{ lb})^2 + (5.00 \text{ lb})^2} = \sqrt{75.00 \text{ lb}^2 + 25.00 \text{ lb}^2} = \sqrt{100.00 \text{ lb}^2} = 10.0 \text{ lb}$$

as we expect.

VECTOR ADDITION BY COMPONENTS

It is possible to calculate the sum of two or more vectors which are neither parallel nor perpendicular by using their components. The procedure is as follows:

1. Resolve the initial vectors into their components in the x (vertical) and y (horizontal) directions.

2. Add the components in the x direction together to give $\mathbf{R}_x$ and add the components in the y direction to give $\mathbf{R}_y$. If the vectors are $\mathbf{A}, \mathbf{B}, \mathbf{C}, \ldots$, the magnitudes of $\mathbf{R}_x$ and $\mathbf{R}_y$ are given by

$$R_x = A_x + B_x + C_x + \cdots \qquad R_y = A_y + B_y + C_y + \cdots$$

3. Calculate the magnitude of the resultant $\mathbf{R}$ by the Pythagorean theorem:

$$R = \sqrt{R_x^2 + R_y^2}$$

The direction of $\mathbf{R}$ can be found by trigonometry.

Problem 11.23. A boat is headed north at a speed of 8.0 mi/h. A strong wind is blowing whose pressure on the boat's superstructure causes it to move sideways to the west at a speed of 2.0 mi/h. There is also a tidal current present that flows in a direction 30° south of east at a speed of 5.0 mi/h. What is the boat's velocity relative to the earth's surface?

The first step is to establish a suitable set of coordinate axes, such as the ones shown in Fig. 11-13(a). Next we draw the three velocity vectors **A**, **B**, and **C**, and calculate the magnitudes of their x and y components. We

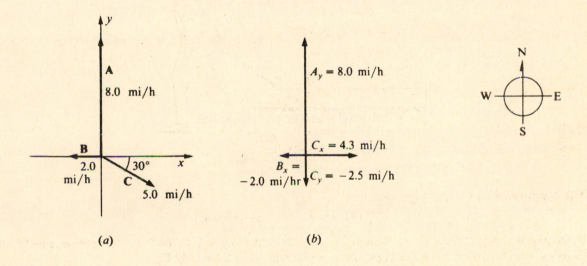

(a)   (b)

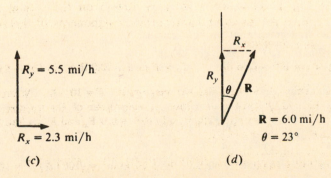

(c)   (d)

Fig. 11-13

find these values:

x components	y components
$A_x = 0$	$A_y = 8.0$ mi/h
$B_x = -2.0$ mi/h	$B_y = 0$
$C_x = C \cos 30°$	$C_y = -C \sin 30°$
$\quad = 5.0$ mi/h $\times 0.866$	$\quad = -5.0$ mi/h $\times 0.500$
$\quad = 4.3$ mi/h	$\quad = -2.5$ mi/h

These components are shown in Fig. 11-13(b).

Now we add up the values of the x components to get R_x and add up the values of the y components to get R_y:

$$R_x = A_x + B_x + C_x = 0 - 2.0 \text{ mi/h} + 4.3 \text{ mi/h} = 2.3 \text{ mi/h}$$
$$R_y = A_y + B_y + C_y = 8.0 \text{ mi/h} + 0 - 2.5 \text{ mi/h} = 5.5 \text{ mi/h}$$

The magnitude of the resultant **R** is therefore

$$R = \sqrt{R_x^2 + R_y^2} = \sqrt{(2.3 \text{ mi/h})^2 + (5.5 \text{ mi/h})^2} = \sqrt{35.54} \text{ mi/h} = 6.0 \text{ mi/h}$$

The boat's speed relative to the earth's surface is 6.0 mi/h.

The direction in which the boat is moving relative to the earth's surface can be given in terms of the angle θ between **R** and the $+y$ axis, which is north.   Since

$$\tan \theta = \frac{R_x}{R_y} = \frac{2.3 \text{ mi/h}}{5.5 \text{ mi/h}} = 0.418 \qquad \theta = 23°$$

the boat's direction of motion is actually 23° to the east of north even though it is headed north [Fig. 11-13(d)].

Supplementary Problems

11.24. Express the following angles in radians:

  (a) 23.5°   (b) 80.4°   (c) 120°   (d) 300°   (e) $3\frac{1}{2}$ rotations

11.25. Express the following angles in degrees:

  (a) 0.1 rad   (b) 0.5 rad   (c) 4.3 rad   (d) $\pi/3$ rad   (e) 4π rad

11.26. Express the following angles in decimal form:

  (a) 5°6′12″   (b) 58°37′40″   (c) 140°20′30″   (d) 225°50′55″   (e) 352°15′25″

11.27. Express the following angles in degrees, minutes, and seconds:

  (a) 1.75°   (b) 15.63°   (c) 70.81°   (d) 164.55°   (e) 320.28°

11.28. Perform the following calculations:

  (a)   43°12′55″ + 12°20′15″   (e)   71°25′32″ × 3

  (b)   70°43′25″ + 30°51′40″   (f)   5°40′12″ × 10

  (c)   65°12′30″ − 12°40′14″   (g)   25°/4

  (d)   160°40′20″ − 30°55′55″   (h)   58°12′40″/5

11.29. Find the following angles, which lie between 0° and 90°:

    (a) $\sin^{-1} 0.720$ (d) $\cos^{-1} 1.000$

    (b) $\cos^{-1} 0.000$ (e) $\tan^{-1} 0.500$

    (c) $\cos^{-1} 0.254$ (f) $\tan^{-1} 8.500$

11.30. Evaluate the following:

    (a) $\sin 100°$ (c) $\sin 300°$ (e) $\cos 150°$ (g) $\tan 170°$ (i) $\tan 190°$

    (b) $\sin 200°$ (d) $\sin 360°$ (f) $\cos 270°$ (h) $\tan 180°$ (j) $\tan 270°$

11.31. Find the length of the hypotenuse of a right triangle whose short sides are 483 ft and 620 ft long.

11.32. The hypotenuse of a right triangle is 28 cm long and the length of one of the short sides is 23 cm long. Find the length of the other side.

11.33. Find the values of the unknown sides and angles in the right triangles for which the following data are known (see Fig. 11-5):

    (a) $\theta = 45°$, $a = 10$ (d) $a = 3$, $b = 4$, $c = 5$

    (b) $\theta = 15°$, $b = 4$ (e) $a = 5$, $b = 12$, $c = 13$

    (c) $\theta = 25°$, $c = 5$

11.34. Two forces, one of 10 lb and the other of 6 lb, act on a body. The directions of the forces are not known. (a) What is the minimum magnitude of the resultant of these forces? (b) What is the maximum magnitude?

11.35. A man drives 10 mi to the north and then 20 mi to the east. What is the magnitude and direction of his displacement from the starting point?

11.36. Find the magnitude and direction of the resultant force produced by a vertically upward force of 40 lb and a horizontal force of 30 lb.

11.37. Find the vertical and horizontal components of a 50-lb force that is directed 50° above the horizontal.

11.38. A man pushes a lawnmower with a force of 20 lb. If the handle of the lawnmower is 40° above the horizontal, how much downward force is being exerted on the ground?

11.39. An airplane is heading northeast (45° east of north) at a speed of 550 mi/h. What is the northward component of its velocity? The eastward component?

11.40. A boat heads northwest at 10 mi/h in a river that flows east at 3 mi/h. What is the magnitude and direction of the boat's velocity relative to the earth's surface?

11.41. A boat moving at 12 mi/h is crossing a river in which the current is flowing at 4 mi/h. In what direction should the boat head if it is to reach a point on the other side of the river directly opposite its starting point?

11.42. An airplane flies 400 mi west from city A to city B, then 300 mi northeast (45° east of north) to city C, and finally 100 mi north to city D. How far is it from city A to city D? In what direction must the airplane head to return directly to city A from city D?

11.43. Find the magnitude and direction of the resultant of a 5-lb force that acts at an angle of 37° clockwise from the $+x$ axis, a 3-lb force that acts at an angle of 180° clockwise from the $+x$ axis, and a 7-lb force that acts at an angle of 225° clockwise from the $+x$ axis.

11.44. Find the magnitude and direction of the resultant of a 60-lb force that acts at an angle of 45° clockwise from the $+y$ axis, a 20-lb force that acts at an angle of 90° clockwise from the $+y$ axis, and a 40-lb force that acts at an angle of 300° clockwise from the $+y$ axis.

Answers to Supplementary Problems

11.24. (*a*) 0.41 rad   (*b*) 1.4 rad   (*c*) 2.09 rad   (*d*) 5.24 rad   (*e*) 22 rad

11.25. (*a*) 5.73°   (*b*) 28.7°   (*c*) 246.4°   (*d*) 60°   (*e*) 720°

11.26. (*a*) 5.103°   (*b*) 58.628°   (*c*) 140.342°   (*d*) 225.849°   (*e*) 352.257°

11.27. (*a*) 1°45′0″   (*b*) 15°37′48″   (*c*) 70°48′36″   (*d*) 164°33′0″   (*e*) 320°16′48″

11.28. (*a*)   55°33′10″   (*c*)   52°32′16″   (*e*)   214°16′36″   (*g*)   6°15′0″
      (*b*)   101°35′5″   (*d*)   129°44′25″   (*f*)   56°42′0″   (*h*)   11°38′32″

11.29. (*a*) 46.1°   (*b*) 90°   (*c*) 75.3°   (*d*) 0°   (*e*) 26.6°   (*f*) 83.3°

11.30. (*a*)   0.985   (*c*)   − 0.866   (*e*)   − 0.866   (*g*)   − 0.176   (*i*)   0.176
      (*b*)   − 0.342   (*d*)   0   (*f*)   0   (*h*)   0   (*j*)   ∞ (infinity)

11.31. 786 ft

11.32. 16 cm

11.33. (*a*) $b = 10, c = 14.1, \phi = 45°$ (*d*) $\theta = 37°, \phi = 53°$
      (*b*) $a = 1.07, c = 4.14, \phi = 75°$ (*e*) $\theta = 23°, \phi = 67°$
      (*c*) $a = 2.11, b = 4.53, \phi = 65°$

11.34. (*a*) 4 lb   (*b*) 16 lb

11.35. 22 mi at 63° east of north

11.36. 50 lb at 53° above the horizontal

11.37. 38.3 lb; 32.1 lb

11.38. 12.9 lb

11.39. 389 mi/h; 389 mi/h

11.40. 8.2 mi/h at 30° west of north

11.41. 19° upstream of directly across the river

11.42. 364 mi at 31° east of south

11.43. 4.4 lb at 206° clockwise from the $+x$ axis

11.44. 68 lb at 24° clockwise from the $+y$ axis

Chapter 12

Alternating Current

GRAPHS

A graph is a pictorial way to display a relationship between two quantities. If we apply a series of different voltages across a 5-Ω resistor and measure the resulting currents, the results would be these:

Voltage, V	0	10	20	30	40	50
Current, A	0	2	4	6	8	10

These measurements are plotted in Fig. 12-1 with the horizontal scale representing voltage and the vertical scale representing current. (The divisions used on each scale can be anything convenient for the numbers involved.) Each point on the graph corresponds to one pair of measurements, and the solid line drawn through the various points shows how I varies with V.

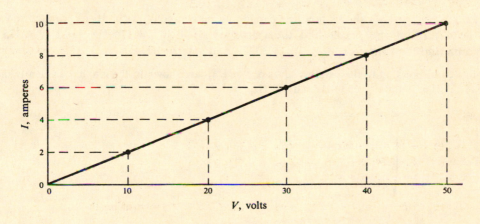

Fig. 12-1

Although it is usually valid to draw a smooth curve to join known points on a graph, a procedure called *interpolation*, situations exist in which the actual curves have irregularities that may not be evident unless many points are established. And it is seldom wise to continue a curve beyond the extreme known points, a procedure called *extrapolation*.

The graph of Fig. 12-1 provides a handy way to find without calculation the current that corresponds to a voltage different from the original ones. If we want to know the current that would correspond to a voltage of 35 V, for instance, we follow the procedure shown in Fig. 12-2:

1. Find the point that represents 35 V on the horizontal scale and draw a vertical line from there to the curve.
2. Draw a horizontal line from the intersection to the vertical scale and read off the current there, which is 7 A.

If we instead want to know the voltage required to yield a current of, say, 5 A, we reverse the

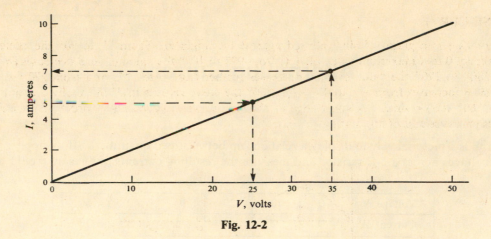

Fig. 12-2

process:

1. Find the point that represents 5 A on the vertical scale and draw a horizontal line from there to the curve.

2. Draw a vertical line from the intersection to the horizontal scale and read off the voltage there, which is 25 V.

Problem 12.1. Figure 12-3 shows how the capacitive reactance X_C of a 10-μF capacitor varies with frequency f. (*a*) From the graph find the values of X_C for $f = 125$ Hz and $f = 375$ Hz. (*b*) From the graph find the values of f for $X_C = 30\ \Omega$ and $X_C = 100\ \Omega$.

The points corresponding to the various values of f and X_C are shown in the graph. Evidently the required values are (*a*) 127 Ω and 42 Ω, (*b*) 530 Hz and 159 Hz.

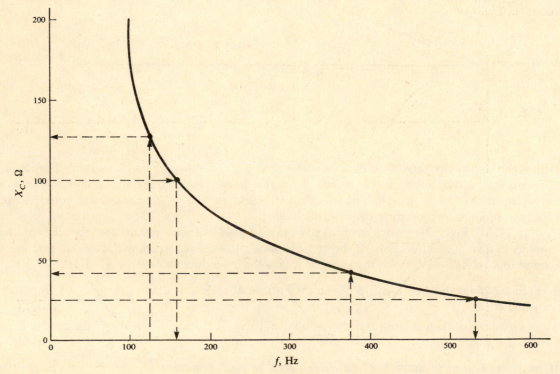

Fig. 12-3

THE SINE CURVE

Figure 12-4 is a graph of $\sin\theta$ plotted against the angle θ. From 0° to 90° the value of $\sin\theta$ increases from 0 to a maximum value of $+1$; from 90° to 180° the value of $\sin\theta$ decreases from $+1$ to 0; from 180° to 270° the value of $\sin\theta$ decreases further from 0 to -1; and from 270° to 360° the value of $\sin\theta$ increases from -1 to 0. After 360° the curve repeats itself in exactly the same way, so that $\sin(360° + \theta) = \sin\theta$. A knowledge of the sine curve is necessary in order to understand the properties of waves and of alternating currents.

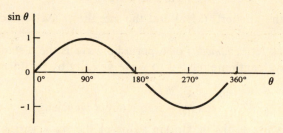

Fig. 12-4

WAVES

A wave is, in general, a disturbance that moves through a medium. A wave carries energy, but there is no transport of matter. In a periodic wave, pulses of the same kind follow one another in regular succession. Sound waves, water waves, and electromagnetic waves are almost always periodic waves whose variations correspond to those of the sine curve: the quantity that changes with time as the wave passes a certain place (pressure in the case of sound waves, water level in the case of water waves, electric and magnetic fields in the case of electromagnetic waves) does so in the manner shown in Fig. 12-5(a). At a particular time, the wave quantity changes with distance in a similar way, as in Fig. 12-5(b). The shapes of the curves of Fig. 12-5(a) and 12-5(b) are the same as the shape of the sine curve of Fig. 12-4.

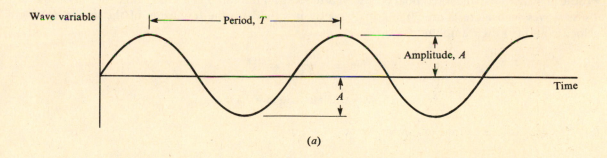

(a)

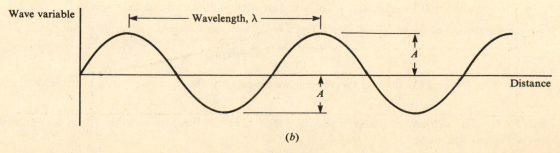

(b)

Fig. 12-5 ·

The *period T* of a wave is the time required for one complete wave to pass a given point. The *frequency f* is the number of waves that pass that point per second, so that

$$f = \frac{1}{T}$$

$$\text{Frequency} = \frac{1}{\text{period}}$$

The unit of frequency is the *hertz* (Hz), where

$$1 \text{ Hz} = 1 \text{ wave/s}$$

Multiples of the hertz are the kilohertz (kHz) and the megahertz (MHz), equal respectively to 10^3 Hz and 10^6 Hz.

The *wavelength λ* (Greek letter *lambda*) of a periodic wave is the distance between adjacent wave crests. Frequency and wavelength are related to wave velocity by the formula

$$v = f\lambda$$

$$\text{Wave velocity} = \text{frequency} \times \text{wavelength}$$

The *amplitude A* of a wave is the maximum value of the wave variable. As mentioned, this is pressure in the case of sound waves, water level in the case of water waves, and electric and magnetic fields in the case of electromagnetic waves.

Electromagnetic waves consist of coupled electric and magnetic fields that vary periodically as they move through space. Radio waves, light waves, x-rays, and gamma rays are examples of electromagnetic waves, and differ only in frequency. The color of light waves depends upon their frequency, with red light having the lowest visible frequencies and violet light the highest. White light is a mixture of light waves of all frequencies.

Electromagnetic waves are generated by accelerated electric charges, usually electrons. Electrons oscillating back and forth in an antenna give off radio waves, for instance, and accelerated electrons in atoms give off light waves. In free space all electromagnetic waves have the *velocity of light*, which is

$$\text{Velocity of light} = c = 3.00 \times 10^8 \text{ m/s}$$

In more familiar units, $c = 186,000$ miles/s.

Problem 12.2. An anchored boat is observed to rise and fall through a total range of 2 m once every 4 s as waves whose crests are 30 m apart pass by it. Find (*a*) the frequency of the waves, (*b*) their velocity, and (*c*) their amplitude.

(*a*)
$$f = \frac{1}{T} = \frac{1}{4 \text{ s}} = 0.25 \text{ Hz}$$

(*b*)
$$v = f\lambda = 0.25 \text{ Hz} \times 30 \text{ m} = 7.5 \text{ m/s}$$

(*c*) The amplitude is half the total range, hence $A = 1$ m.

Problem 12.3. As a phonograph record turns, a certain groove passes the needle at 25 cm/s. If the wiggles in the groove are 0.1 mm apart, what is the frequency of the sound that results?

Here the wavelength of the wiggles is $\lambda = 0.1$ mm $= 10^{-4}$ m, so they pass the needle at the rate of

$$f = \frac{v}{\lambda} = \frac{0.25 \text{ m/s}}{10^{-4} \text{ m}} = 2500 \text{ Hz}$$

This is therefore the frequency of the sound waves that are produced by the electronic system of the record player.

Problem 12.4. The velocity of sound in seawater is 5020 ft/s. Find the wavelength in seawater of a sound wave whose frequency is 256 Hz.

$$\lambda = \frac{v}{f} = \frac{5020 \text{ ft/s}}{256 \text{ Hz}} = 19.6 \text{ ft}$$

Problem 12.5.   A marine radar operates at a wavelength of 3.2 cm.   What is the frequency of the radar waves?

Radar waves are electromagnetic and hence travel with the velocity of light c.   Therefore

$$f = \frac{c}{\lambda} = \frac{3 \times 10^8 \text{ m/s}}{3.2 \times 10^{-2} \text{ m}} = 9.4 \times 10^9 \text{ Hz}$$

ALTERNATING CURRENT

In a generator, a coil of wire is rotated in a magnetic field to produce an electric current by electromagnetic induction.   In this way mechanical energy is converted to electric energy.

When a wire loop is rotated about a diameter in a magnetic field, each half of the loop moves up and then down through the field.   As a result the current induced in the loop constantly changes in magnitude and reverses its direction twice per complete turn.   Such a changing current is called an *alternating current* (ac), and all generators produce it.   If direct (one-way) current is required, either the output of the generator can be switched back and forth automatically as the coil turns by means of a commutator and brushes, or an external device called a *rectifier* can be used which permits current to pass through it in only one direction.   Both approaches are widely used.

Figure 12-6 shows how the output voltage V of an ac generator varies with time.   The frequency of an alternating current is the number of complete back-and-forth cycles it goes through each second. The unit of frequency, as mentioned above, is the hertz (Hz), where   1 Hz = 1 cycle/s.   The curve of Fig. 12-6 has exactly the same shape as that of Fig. 12-4 for the sine of an angle.   The way that V varies with time t is given by the formula

$$V = V_{max} \sin 2\pi f t = V_{max} \sin \omega t$$

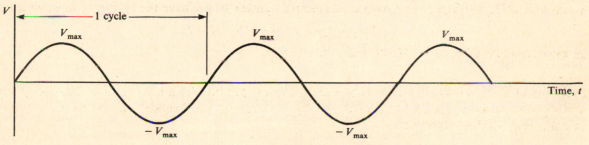

Fig. 12-6

In this formula V_{max} is the maximum value of the voltage, corresponding to the amplitude of a wave. The quantity $\omega = 2\pi f$ is called the *angular frequency* of the alternating current; the units of ω are radians/second.   (ω is the lower-case Greek letter "omega.")

The current I in an ac circuit varies with time in the same way as the voltage.   If the frequency of the current is f and the maximum value of the current is I_{max}, then

$$I = I_{max} \sin 2\pi f t = I_{max} \sin \omega t$$

EFFECTIVE VALUES

Because an alternating current changes continuously, its maximum value $\pm I_{max}$ does not indicate its ability to do work or to produce heat as does the magnitude of a direct current.   Instead it is customary to refer to the *effective current*

$$I_{eff} = \frac{I_{max}}{\sqrt{2}} = 0.707 I_{max}$$

which is such that a direct current of this value does as much work or produces as much heat as the alternating current whose maximum value is $\pm I_{max}$. Similarly the effective voltage in an ac circuit is

$$V_{eff} = \frac{V_{max}}{\sqrt{2}} = 0.707 V_{max}$$

Currents and voltages in ac circuits are usually expressed in terms of their effective values, which are 70.7 percent of their maximum values. The values of I and V at any particular moment are called the *instantaneous* values of these quantities.

The effective values of I and V are sometimes called "root-mean-square" or "rms" values because of the way they are defined.

Problem 12.6. Find the maximum voltage across a 120-V ac power line.

$$\pm V_{max} = \frac{\pm V_{eff}}{0.707} = \frac{\pm 120 \text{ V}}{0.707} = \pm 170 \text{ V}$$

Problem 12.7. The dielectric used in a certain capacitor breaks down at a potential difference of 300 V. Find the maximum effective ac potential difference that can be applied to it.

$$V_{eff} = 0.707 V_{max} = 0.707 \times 300 \text{ V} = 212 \text{ V}$$

Problem 12.8. Alternating current with a maximum value of 10 A is passed through a 20-Ω resistor. At what rate does the resistor dissipate energy?

The effective current is

$$I_{eff} = 0.707 I_{max} = 0.707 \times 10 \text{A} = 7.07 \text{ A}$$

and so the power dissipated is

$$P = I_{eff}^2 R = (7.07 \text{ A})^2 \times 20 \ \Omega = 1000 \text{ W}$$

THE TRANSFORMER

A *transformer* consists of two coils of wire, usually wound on an iron core. When an alternating current is passed through one of the windings, the changing magnetic field it gives rise to induces an alternating current in the other winding. The potential difference per turn is the same in both primary and secondary windings, so the ratio of turns in the windings determines the ratio of voltages across them:

$$\frac{V_1}{V_2} = \frac{N_1}{N_2}$$

$$\frac{\text{Primary voltage}}{\text{Secondary voltage}} = \frac{\text{primary turns}}{\text{secondary turns}}$$

Since the power $I_1 V_1$ going into a transformer must equal the power $I_2 V_2$ going out, where I_1 and I_2 are the primary and secondary currents respectively, the ratio of currents is inversely proportional to the ratio of turns:

$$\frac{I_1}{I_2} = \frac{N_2}{N_1}$$

$$\frac{\text{Primary current}}{\text{Secondary current}} = \frac{\text{secondary turns}}{\text{primary turns}}$$

Problem 12.9. A transformer has 100 turns in its primary winding and 500 turns in its secondary winding. If the primary voltage and current are respectively 120 V and 3 A, what are the secondary voltage and current?

$$V_2 = \frac{N_2}{N_1} V_1 = \frac{500 \text{ turns}}{100 \text{ turns}} \times 120 \text{ V} = 600 \text{ V}$$

$$I_2 = \frac{N_1}{N_2} I_1 = \frac{100 \text{ turns}}{500 \text{ turns}} \times 3 \text{ A} = 0.6 \text{ A}$$

Problem 12.10. A transformer rated at a maximum power of 10 kW is used to connect a 5000-V transmission line to a 240-V circuit. (*a*) What is the ratio of turns in the windings of the transformer? (*b*) What is the maximum current in the 240-V circuit?

(*a*)
$$\frac{N_1}{N_2} = \frac{V_1}{V_2} = \frac{5000 \text{ V}}{240 \text{ V}} = 20.8$$

(*b*) Since $P = IV$ here,

$$I_2 = \frac{P}{V_2} = \frac{10\,000 \text{ W}}{240 \text{ V}} = 41.7 \text{ A}$$

Problem 12.11. A transformer connected to a 120-V ac power line has 200 turns in its primary winding and 50 turns in its secondary winding. The secondary is connected to a 100-Ω light bulb. How much current is drawn from the 120-V power line?

The voltage across the secondary is

$$V_2 = \frac{N_2}{N_1} V_1 = \frac{50 \text{ turns}}{200 \text{ turns}} \times 120 \text{ V} = 30 \text{ V}$$

and so the current in the secondary circuit is

$$I_2 = \frac{V_2}{R} = \frac{30 \text{ V}}{100 \ \Omega} = 0.3 \text{ A}$$

Hence the current in the primary circuit is

$$I_1 = \frac{N_2}{N_1} I_2 = \frac{50 \text{ turns}}{200 \text{ turns}} \times 0.3 \text{ A} = 0.075 \text{ A}$$

Problem 12.12. Alternating current is in wide use chiefly because its voltage can be so easily changed by a transformer. Since $P = IV$, the higher the voltage, the lower the current, and vice versa. In transmitting electric energy through long distances, a small current is desirable in order to minimize energy loss to heat, which is equal to I^2R where R is the resistance of the transmission line. On the other hand, both the generation and final use of electric energy are best accomplished at moderate potential differences. Hence electricity is typically generated at 10 000 V or so, stepped up by transformers at the power station to 500 000 V or even more for transmission, and near the point of consumption other transformers reduce the potential difference to 240 V or 120 V. To verify the advantage of high-voltage transmission, find the rate of energy loss to heat when a 5-Ω cable is used to transmit 1 kW of electricity at 100 V and at 100 000 V.

Since $P = IV$, the currents in the cable are respectively

$$I_A = \frac{P}{V_A} = \frac{1000 \text{ W}}{100 \text{ V}} = 10 \text{ A} \qquad I_B = \frac{P}{V_B} = \frac{1000 \text{ W}}{100\,000 \text{ V}} = 0.01 \text{ A}$$

The rates of heat production per kilowatt are respectively

$$I_A^2 R = (10 \text{ A})^2 \times 5 \ \Omega = 500 \text{ W}$$

$$I_B^2 R = (0.01 \text{ A})^2 \times 5 \ \Omega = 0.0005 \text{ W} = 5 \times 10^{-4} \text{ W}$$

Transmission at 100 V therefore means 10^6—a million—times more energy lost as heat than does transmission at 100 000 V.

PHASOR REPRESENTATION

A convenient way to represent a quantity that varies sinusoidally (that is, in the way that $\sin \theta$ varies with θ) with time is in terms of a rotating vector called a *phasor*. In the case of an ac voltage, the length of the phasor $\mathbf{V}_{max}$ corresponds to V_{max} and we imagine it to rotate f times per second in a counterclockwise direction (Fig. 12-7). The vertical component of the phasor at any moment corresponds to the instantaneous voltage V. Since the vertical component of V_{max} is given by

$$V = V_{max} \sin \theta = V_{max} \sin 2\pi ft$$

the result is the same curve as that of Fig. 12-6. Figure 12-8 shows how a rotating phasor generates a sine curve. In the next chapters we will find phasors very helpful.

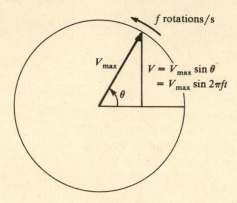

Fig. 12-7

Problem 12.13. Find the value of V when $V_{max} = 100$ V for $\theta = 50°$, $120°$, $260°$, and $305°$.

$V_1 = 100$ V $\times \sin 50° = 100$ V $\times 0.766 = 76.6$ V

$V_2 = 100$ V $\times \sin 120° = 100$ V $\times \sin(90° + 30°) = 100$ V $\times \cos 30° = 100$ V $\times 0.866 = 86.6$ V

$V_3 = 100$ V $\times \sin 260° = 100$ V $\times \sin(180° + 80°) = 100$ V $\times (-\sin 80°) = 100$ V $\times (-0.985) = -98.5$ V

$V_4 = 100$ V $\times \sin 305° = 100$ V $\times \sin(270° + 35°) = 100$ V $\times (-\cos 35°) = 100$ V $\times (-0.819) = -81.9$ V

Supplementary Problems

12.14. The velocity of sound waves in air at sea level is 331 m/s. Find the wavelength in air of a sound wave whose frequency is 440 Hz.

12.15. A tuning fork vibrating at 600 Hz is immersed in a tank of water, and the resulting sound waves in the water are found to have a wavelength of 8.2 ft. What is the velocity of sound in water?

12.16. What is the frequency of radio waves whose wavelength is 20 m? (Note that the velocity of radio waves is the same as the velocity of light.)

12.17. A certain radio station broadcasts at 1050 kHz. What is the wavelength of these waves?

12.18. A voltmeter across an ac circuit reads 40 V and an ammeter in series with a circuit reads 6 A. (*a*) What is the maximum voltage across the circuit? (*b*) What is the maximum current in the circuit?

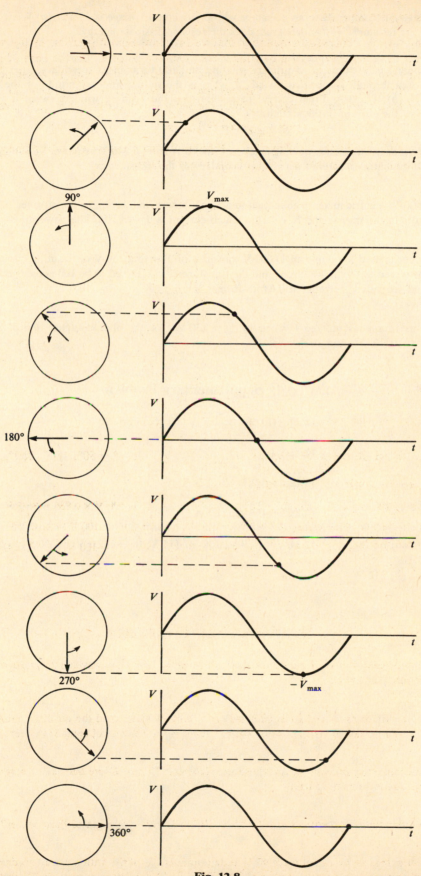

Fig. 12-8

12.19. A 25-Ω resistor is connected to an ac source whose maximum voltage is 75 V. (*a*) What is the effective current in the resistor? (*b*) At what rate does the resistor dissipate energy?

12.20. A transformer has 50 turns in its primary winding and 100 turns in its secondary. (*a*) If a 60-Hz, 3-A current passes through the primary winding, what is the nature and magnitude of the current in the secondary? (*b*) If a 3-A direct current passes through the primary winding, what is the nature and magnitude of the current in the secondary?

12.21. The transformer in an electric welding machine draws 3 A from a 240-V ac power line and delivers 400 A. What is the potential difference across the secondary of the transformer?

12.22. A 240-V, 400-W electric mixer is connected to a 120-V power line through a transformer. (*a*) What is the ratio of turns in the transformer? (*b*) How much current is drawn from the power line?

12.23. A transformer is used to couple an 11 000-V transmission line to a 230-V distribution circuit. If 5 kW of power is delivered to the 230-V circuit and the transformer is assumed to be 100 percent efficient, find the primary and secondary currents in the transformer.

12.24. Find the instantaneous voltage V when $V_{max} = 230$ V for the phasor angles $\theta = 0$, 10°, 90°, 150°, 180°, 220°, 270°, 345°, and 360°.

Answers to Supplementary Problems

12.14. 0.75 m

12.15. 4920 ft/s

12.16. 15 MHz

12.17. 286 m

12.18. (*a*) 56.6 V   (*b*) 8.5 A

12.19. (*a*) 2.12 A   (*b*) 112 W

12.20. (*a*) 60 Hz, 1.5 A   (*b*) No current

12.21. 1.8 V

12.22. (*a*) 2:1   (*b*) 3.3 A

12.23. 0.455 A; 21.7 A

12.24. 0; 39.9 V; 230 V; 115 V; 0; −147.8 V; −230 V; −59.5 V; 0

Chapter 13

Series AC Circuits

REACTANCE

The *inductive reactance* of an inductor is a measure of its effectiveness in opposing the flow of an alternating current by virtue of the self-induced back emf that the changing current causes in it. Unlike the case of a resistor, there is no power dissipated in a pure inductor. The inductive reactance X_L of an inductor whose inductance is L (in henries) when the frequency of the current is f (in Hz) is

$$\text{Inductive reactance} = X_L = 2\pi f L$$

When a potential difference V of frequency f is applied across the inductor whose reactance is X_L at the frequency f, the current $I = V/X_L$ will flow. The unit of X_L is the ohm.

The *capacitive reactance* of a capacitor is similarly a measure of its effectiveness in opposing the flow of an alternating current, in this case by virtue of the reverse potential difference across it due to the accumulation of charge on its plates. No power loss is associated with a capacitor in an ac circuit. The capacitive reactance X_C of a capacitor whose capacitance is C (in farads) when the frequency of the current is f (in Hz) is

$$\text{Capacitive reactance} = X_C = \frac{1}{2\pi f C}$$

When a potential difference of frequency f is applied across a capacitor whose reactance is X_C at the frequency f, the current $I = V/X_C$ will flow. The unit of X_C is the ohm.

Problem 13.1. (a) What happens to X_L and X_C in the limit of $f = 0$? (b) What is the physical meaning of these results?

(a) When $f = 0$, $X_L = 2\pi f L = 0$ and $X_C = 1/2\pi f C = \infty$.

(b) A current with $f = 0$ is a direct current. When a constant current flows in an inductor, there is no self-induced back emf to hamper the current, and the inductive reactance is accordingly 0. A direct current cannot pass through a capacitor because its plates are insulated from each other, so the capacitive reactance is infinite and $I = V/X_C = 0$ when $f = 0$. (An alternating current does not actually pass *through* a capacitor but surges back and forth in the circuit on both sides of it.)

Problem 13.2. A 10-μF capacitor is connected to a 15-V, 5-kHz power source. Find (a) the reactance of the capacitor and (b) the current that flows.

(a)
$$X_C = \frac{1}{2\pi f C} = \frac{1}{2\pi \times 5 \times 10^3 \text{ Hz} \times 10 \times 10^{-6} \text{ F}} = 3.18 \ \Omega$$

(b)
$$I = \frac{V}{X_C} = \frac{15 \text{ V}}{3.18 \ \Omega} = 4.72 \text{ A}$$

Problem 13.3. The reactance of a capacitor is 50 Ω at 300 Hz. What is its capacitance?

$$X_C = \frac{1}{2\pi f C}$$

Solving for C,

$$C = \frac{1}{2\pi f X_C} = \frac{1}{2\pi \times 300 \text{ Hz} \times 50} = 1.06 \times 10^{-5} \text{ F} = 10.6 \times 10^{-6} \text{ F} = 10.6 \text{ F}$$

127

Problem 13.4. An 0.3-H inductor of negligible resistance is connected to a 24-V, 60-Hz power source.   Find (a) the reactance of the inductor and (b) the current that flows.

(a)
$$X_L = 2\pi fL = 2\pi \times 60 \text{ Hz} \times 0.3 \text{ H} = 113 \ \Omega$$

(b)
$$I = \frac{V}{X_L} = \frac{24 \text{ V}}{113 \ \Omega} = 0.21 \text{ A}$$

Problem 13.5. The reactance of an inductor is 80 Ω at 500 Hz.   Find its inductance.

Since $X_L = 2\pi fL$,

$$L = \frac{X_L}{2\pi f} = \frac{80 \ \Omega}{2\pi \times 500 \text{ Hz}} = 0.0255 \text{ H} = 25.5 \text{ mH}$$

PHASE ANGLE

In an ac circuit that contains only resistance, the instantaneous voltage and current are *in phase* with each other; that is, both are 0 at the same time, both reach their maximum values in either direction at the same time, and so on, as shown in Fig. 13-1(a).

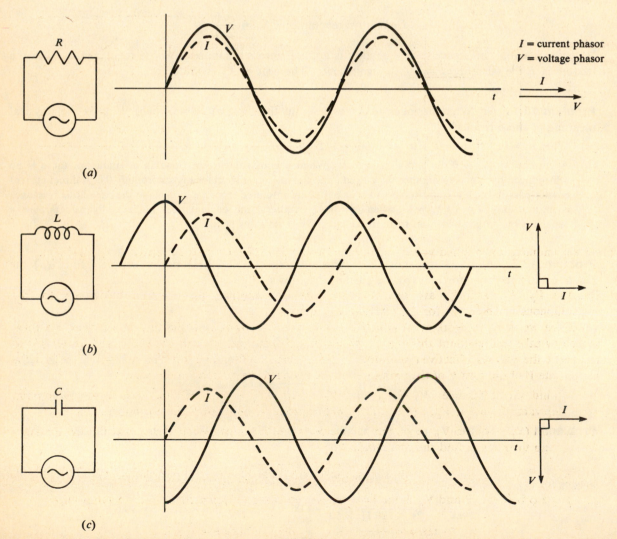

Fig. 13-1

In an ac circuit that contains only inductance, the voltage leads the current by 1/4 cycle. Since a complete cycle means a change in $2\pi ft$ of $360°$ and $360°/4 = 90°$, it is customary to say that in a pure inductor the voltage leads the current by $90°$. This situation is shown in Fig. 13-1(*b*).

In an ac circuit that contains only capacitance, the voltage lags behind the current by 1/4 cycle, which is $90°$. This situation is shown in Fig. 13-1(*c*).

Now we consider an ac circuit that contains resistance, inductance, and capacitance in series, as in Fig. 13-2. The instantaneous voltages across the circuit elements are as follows:

$$V_R = IR \qquad V_L = IX_L \qquad V_C = IX_C$$

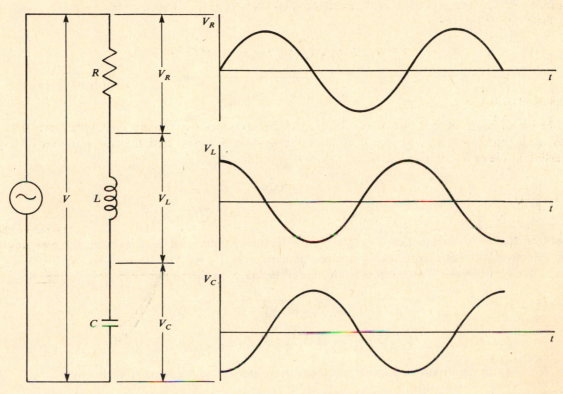

Fig. 13-2

At any moment the applied voltage V is equal to the sum of the voltage drops V_R, V_L, and V_C:

$$V = V_R + V_L + V_C$$

Because V_R, V_L, and V_C are out of phase with one another, this formula holds only for the instantaneous voltages, *not* for the effective voltages.

Since we want to work with effective voltages and currents, not instantaneous ones, we must somehow take into account the phase differences. To do this, we can use phasors (Chapter 12) to represent the various effective quantities. This is done in Fig. 13-3 for the voltages. To find the magnitude V of the sum $\mathbf{V}$ of the various effective voltages, we proceed in this way:

1. Find the difference $\mathbf{V}_L - \mathbf{V}_C$. If $V_L > V_C$, $(\mathbf{V}_L - \mathbf{V}_C)$ will be positive and will point upward; if $V_L < V_C$, $(\mathbf{V}_L - \mathbf{V}_C)$ will be negative and will point downward.

2. Add $(\mathbf{V}_L - \mathbf{V}_C)$ to $\mathbf{V}_R$ to obtain $\mathbf{V}$. Since $(\mathbf{V}_L - \mathbf{V}_C)$ is perpendicular to $\mathbf{V}_R$, use the Pythagorean theorem to find the magnitude V:

$$V = \sqrt{V_R^2 + (V_L - V_C)^2}$$

The angle ϕ between $\mathbf{V}$ and $\mathbf{V}_R$ is the *phase angle* and can be calculated from the relationships

$$\tan \phi = \frac{V_L - V_C}{V_R} \qquad \text{or} \qquad \cos \phi = \frac{V_R}{V}$$

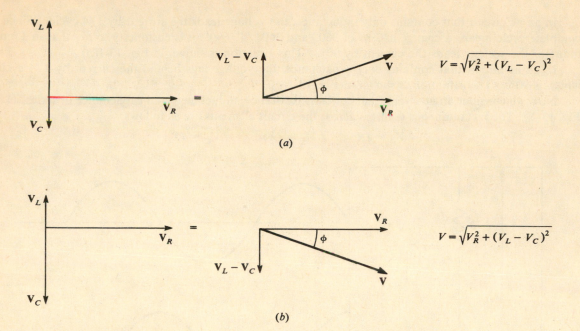

Fig. 13-3

Problem 13.6. A resistor, a capacitor, and an inductor are connected in series to an ac power source. The effective voltages across the circuit components are $V_R = 5$ V, $V_C = 10$ V, and $V_L = 7$ V. Find the effective voltage of the source and the phase angle in the circuit.

(a)
$$V = \sqrt{V_R^2 + (V_L - V_C)^2} = \sqrt{(5 \text{ V})^2 + (7 \text{ V} - 10 \text{ V})^2}$$

$$= \sqrt{(5 \text{ V})^2 + (-3 \text{ V})^2} = \sqrt{25 \text{ V}^2 + 9 \text{ V}^2} = \sqrt{34 \text{ V}^2} = 5.8 \text{ V}$$

We note that the effective voltages across C and R are greater than the effective applied voltage.

(b)
$$\tan \phi = \frac{V_L - V_C}{V_R} = \frac{7 \text{ V} - 10 \text{ V}}{5 \text{ V}} = -\frac{3}{5} = -0.6$$

$$\phi = -31°$$

The negative phase angle means that the voltage across the resistor is ahead of the applied voltage, as in Fig. 13-3(b). Equivalently we can say that the current in the circuit leads the voltage, as in Fig. 13-1(c). (We recall from Chapter 12 that phasors rotate counterclockwise.)

IMPEDANCE

Because

$$V_R = IR \qquad V_L = IX_L \qquad V_C = IX_C$$

we can rewrite the above formula for V in the form

$$V = I\sqrt{R^2 + (X_L - X_C)^2}$$

The quantity

$$Z = \sqrt{R^2 + (X_L - X_C)^2}$$

is called the *impedance* of the circuit and corresponds to the resistance of a dc circuit. The unit of Z is the ohm. When an ac voltage whose frequency is f is applied to a circuit whose impedance is Z at that frequency, the result is the current

$$I = \frac{V}{Z}$$

Figure 13-4 shows phasor impedance diagrams that correspond to the phasor voltage diagrams of Fig. 13-3.   The phase angle ϕ can be calculated from either of these formulas:

$$\tan \phi = \frac{X_L - X_C}{R} \qquad \text{or} \qquad \cos \phi = \frac{R}{Z}$$

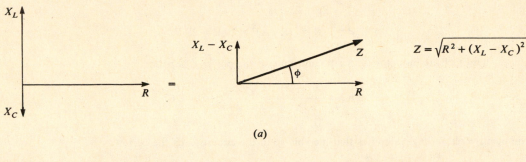

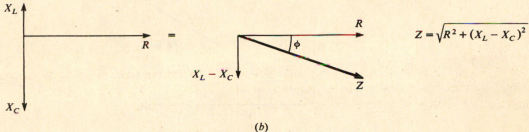

(b)

Fig. 13-4

Problem 13.7. A 5-μF capacitor is in series with a 300-Ω resistor and a 120-V, 60-Hz voltage is applied to the combination.   Find (a) the current in the circuit, (b) the power dissipated in it, and (c) the phase angle.

(a) The reactance of the capacitor at 60 Hz is

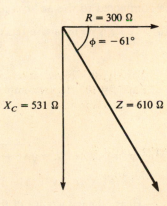

$$X_C = \frac{1}{2\pi f C} = \frac{1}{2\pi \times 60 \text{ Hz} \times 5 \times 10^{-6} \text{ F}} = 531 \ \Omega$$

Since $X_L = 0$,   the impedance of the circuit is (Fig. 13-5)

$$Z = \sqrt{R^2 + (X_L - X_C)^2} = \sqrt{R^2 + (-X_C)^2}$$

$$= \sqrt{R^2 + X_C^2} = \sqrt{(300 \ \Omega)^2 + (531 \ \Omega)^2} = 610 \ \Omega$$

Hence the current is

$$I = \frac{V}{Z} = \frac{120 \text{ V}}{610 \ \Omega} = 0.197 \text{ A}$$

(b) Power is dissipated only in the resistor, so

Fig. 13-5

$$P = I^2 R = (0.197 \text{ A})^2 \times 300 \ \Omega = 11.6 \text{ W}$$

The reactance of the capacitor does not contribute to the power dissipation even though it must be taken into account in determining the current in the circuit.

(c)

$$\tan \phi = \frac{X_L - X_C}{R} = \frac{0 - 531}{300} = -1.77$$

$$\phi = -61°$$

The negative phase angle signifies that the current in the circuit leads the voltage.

Problem 13.8. A 5-mH, 20-Ω inductor is connected to a 28-V, 400-Hz power source. Find (a) the current in the inductor, (b) the power dissipated in it, and (c) the phase angle.

(a) The reactance of the inductor is

$$X_L = 2\pi f L = 2\pi \times 400 \text{ Hz} \times 5 \times 10^{-3} \text{ H} = 12.6 \ \Omega$$

and its impedance is (Fig. 13-6)

$$Z = \sqrt{R^2 + X_L^2} = \sqrt{(20 \ \Omega)^2 + (12.6 \ \Omega)^2} = 23.6 \ \Omega$$

Hence the current is

$$I = \frac{V}{Z} = \frac{28 \text{ V}}{23.6 \ \Omega} = 1.18 \text{ A}$$

(b)

$$P = I^2 R = (1.18 \text{ A})^2 \times 20 \ \Omega = 28 \text{ W}$$

The reactance of the inductor does not contribute to the power dissipation.

(c)

$$\tan \phi = \frac{X_L - X_C}{R} = \frac{12.6 \ \Omega - 0}{20 \ \Omega} = 0.63$$

$$\phi = 32°$$

The positive phase angle signifies that the voltage in the circuit leads the current.

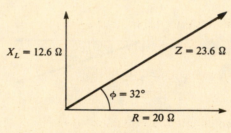

Fig. 13-6

Problem 13.9. A coil of unknown resistance and inductance draws 4 A when connected to a 12-V dc power source and 3 A when connected to a 12-V, 100-Hz power source. (a) Find the values of R and L. (b) How much power is dissipated when the coil is connected to the dc source? (c) When it is connected to the ac source?

(a) There is no inductive reactance when direct current passes through the coil, so its resistance is

$$R = \frac{V_1}{I_1} = \frac{12 \text{ V}}{4 \text{ A}} = 3 \ \Omega$$

At $f = 100$ Hz the impedance of the circuit is

$$Z = \frac{V_2}{I_2} = \frac{12 \text{ V}}{3 \text{ A}} = 4 \ \Omega$$

and so, since $Z = \sqrt{R^2 + (X_L - X_C)^2}$ and $X_C = 0$ here,

$$X_L = \sqrt{Z^2 - R^2} = \sqrt{(4 \ \Omega)^2 - (3 \ \Omega)^2} = 2.65 \ \Omega$$

Hence the inductance of the coil is

$$L = \frac{X_L}{2\pi f} = \frac{2.65 \ \Omega}{2\pi \times 100 \text{ Hz}} = 4.22 \text{ mH}$$

(b)

$$P_1 = I_1^2 R = (4 \text{ A})^2 \times 3 \ \Omega = 48 \text{ W}$$

(c)

$$P_2 = I_2^2 R = (3 \text{ A})^2 \times 3 \ \Omega = 27 \text{ W}$$

Problem 13.10. A 10-μF capacitor, a 0.10-H inductor, and a 60-Ω resistor are connected in series across a 120-V, 60-Hz power source. Find (a) the current in the circuit, (b) the power dissipated in it, and (c) the phase angle.

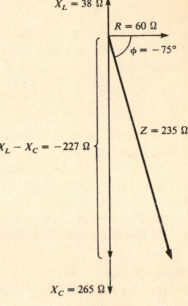

(a) The reactances are

$$X_L = 2\pi fL = 2\pi \times 60 \text{ Hz} \times 0.10 \text{ H} = 38 \text{ }\Omega$$

$$X_C = \frac{1}{2\pi fC} = \frac{1}{2\pi \times 60 \text{ Hz} \times 10 \times 10^{-6} \text{ F}} = 265 \text{ }\Omega$$

The impedance is therefore (Fig. 13-7)

$$Z = \sqrt{R^2 + (X_L - X_C)^2} = \sqrt{(60 \text{ }\Omega)^2 + (38 \text{ }\Omega - 265 \text{ }\Omega)^2} = 235 \text{ }\Omega$$

Hence the current in the circuit is

$$I = \frac{V}{Z} = \frac{120 \text{ V}}{235 \text{ }\Omega} = 0.51 \text{ A}$$

(b) $$P = I^2 R = (0.51 \text{ A})^2 \times 60 \text{ }\Omega = 15.6 \text{ W}$$

(c) $$\tan\phi = \frac{X_L - X_C}{R} = \frac{38 \text{ }\Omega - 265 \text{ }\Omega}{60 \text{ }\Omega} = -\frac{227 \text{ }\Omega}{60 \text{ }\Omega} = -3.78$$

$$\phi = -75°$$

The negative phase angle signifies that the current in the circuit leads the voltage.

Fig. 13-7

RESONANCE

The impedance in a series ac circuit is a minimum when $X_L = X_C$; under these circumstances $Z = R$ and $I = V/R$. The *resonant frequency* f_0 of a circuit is that frequency at which $X_L = X_C$:

$$2\pi f_0 L = \frac{1}{2\pi f_0 C} \qquad f_0 = \frac{1}{2\pi\sqrt{LC}}$$

When the potential difference applied to a circuit has the frequency f_0, the current in the circuit will be a maximum. This condition is known as *resonance*. At resonance the phase angle is 0 since $X_L = X_C$.

Problem 13.11. In the antenna circuit of a radio receiver that is tuned to a particular station, $R = 5 \text{ }\Omega$, $L = 5 \text{ mH}$, and $C = 5 \text{ pF}$. (a) Find the frequency of the station. (b) If the potential difference applied to the circuit is $5 \times 10^{-4} \text{ V}$, find the current that flows.

(a) $$f_0 = \frac{1}{2\pi\sqrt{LC}} = \frac{1}{2\pi\sqrt{5 \times 10^{-3} \text{ H} \times 5 \times 10^{-12} \text{ F}}} = 1007 \text{ kHz}$$

(b) At resonance, $X_L = X_C$ and $Z = R$. Hence

$$I = \frac{V}{R} = \frac{5 \times 10^{-4} \text{ V}}{5 \text{ }\Omega} = 10^{-4} \text{ A} = 0.1 \text{ mA}$$

Problem 13.12. In the antenna circuit of Problem 13.11 the inductance is fixed but the capacitance can be varied. What should the capacitance be in order to receive an 800-kHz radio signal?

Since $2\pi f_0 L = 1/(2\pi f_0 C)$,

$$C = \frac{1}{(2\pi f)^2 L} = \frac{1}{(2\pi \times 800 \times 10^3 \text{ Hz})^2 \times 5 \times 10^{-3} \text{ H}} = 7.9 \times 10^{-12} \text{ F} = 7.9 \text{ pF}$$

POWER FACTOR

The power absorbed in an ac circuit is given by

$$P = IV \cos \phi$$

because, from Fig. 13-3, $V \cos \phi$ is the component of **V** that is in phase with the current $I = V_R/R$.

The quantity $\cos \phi$ is the *power factor* of the circuit. At resonance, $\phi = 0$, $\cos \phi = 1$ and the power absorbed is a maximum. The power factor in an ac circuit is equal to the ratio between its resistance and its impedance:

$$\text{Power factor} = \cos \phi = \frac{R}{Z} = \frac{R}{\sqrt{R^2 + (X_L - X_C)^2}}$$

Power factors are often expressed as percentages, so that a phase angle of, say, 25° would give rise to a power factor of $\cos 25° = 0.906 = 90.6$ percent.

Ac power sources are usually rated in *voltamperes*, the product of V_{eff} and I_{eff}, without regard to the actual power P, because higher values of these quantities must be supplied to a circuit whose power factor is less than 1 than is indicated by its power rating in watts. Thus a power factor of 90.6 percent means that an apparent power of $1 \text{ V} \cdot \text{A}$ (voltampere) must be supplied for each 0.906 W of true power that is consumed by the circuit.

Problem 13.13. A 50-μF capacitor, a 0.3-H inductor, and an 80-Ω resistor are connected in series with a 120-V, 60-Hz power source (Fig. 13.8). (*a*) What is the impedance of the circuit? (*b*) How much current flows in it? (*c*) What is the phase angle? (*d*) How much power is dissipated by the circuit? (*e*) What must be the minimum rating in voltamperes of the power source?

(*a*)
$$X_L = 2\pi f L = 2\pi \times 60 \text{ Hz} \times 0.3 \text{ H} = 113 \ \Omega$$

$$X_C = \frac{1}{2\pi f C} = \frac{1}{2\pi \times 60 \text{ Hz} \times 50 \times 10^{-6} \text{ F}} = 53 \ \Omega$$

$$Z = \sqrt{R^2 + (X_L - X_C)^2} = \sqrt{80^2 + (113 \ \Omega - 53 \ \Omega)^2} = 100 \ \Omega$$

See Fig. 13-9.

(*b*)
$$I = \frac{V}{Z} = \frac{120 \text{ V}}{100 \ \Omega} = 1.2 \text{ A}$$

(*c*)
$$\cos \phi = \frac{R}{Z} = \frac{80 \ \Omega}{100 \ \Omega} = 0.8$$
$$\phi = 37°$$

(*d*)
$$\text{True power} = P = IV \cos \phi = 1.2 \text{ A} \times 120 \text{ V} \times 0.8 = 115 \text{ W}$$

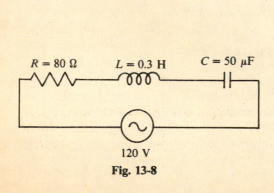

$R = 80 \ \Omega$ $L = 0.3$ H $C = 50 \ \mu$F

120 V

Fig. 13-8

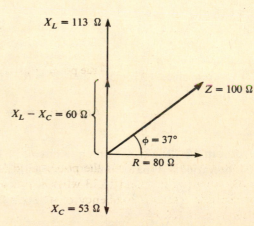

$X_L = 113 \ \Omega$

$Z = 100 \ \Omega$

$X_L - X_C = 60 \ \Omega$

$\phi = 37°$

$R = 80 \ \Omega$

$X_C = 53 \ \Omega$

Fig. 13-9

Alternatively,

$$P = I^2 R = (1.2 \text{ A})^2 \times 80 \ \Omega = 115 \text{ W}$$

(e)   Apparent power $= IV = 1.2 \text{ A} \times 120 \text{ V} = 144 \text{ V} \cdot \text{A}$

Problem 13.14. (a) Find the potential differences across the resistor, the inductor, and the capacitor in the circuit of Problem 13.13. (b) Are these values in accord with the applied potential difference of 120 V?

(a)   Resistor: $V_R = IR = 1.2 \text{ A} \times 80 \ \Omega = 96 \text{ V}$

      Inductor: $V_L = IX_L = 1.2 \text{ A} \times 113 \ \Omega = 136 \text{ V}$

      Capacitor: $V_C = IX_C = 1.2 \text{ A} \times 53 \ \Omega = 64 \text{ V}$

(b)   The sum of these potential differences is 296 V, more than twice the 120 V applied to the circuit. However, this is a meaningless way to combine the potential differences since they are not in phase with one another: V_L is 90° ahead of V_R and V_C is 90° behind V_R. The correct way to find the total potential difference across the circuit is as follows:

$$V = \sqrt{V_R^2 + (V_L - V_C)^2} = \sqrt{(96 \text{ V})^2 + (136 \text{ V} - 64 \text{ V})^2} = 120 \text{ V}$$

This result is in agreement with the applied potential difference of 120 V. We note that the voltage across an inductor or capacitor in an ac circuit can be greater than the voltage applied to the circuit.

Problem 13.15. (a) Find the resonant frequency f_0 of the circuit of Problem 13.13. (b) What current will flow in the circuit if it is connected to a 120-V power source whose frequency is f_0? (c) What will be the power factor in this case? (d) How much power will be dissipated by the circuit? (e) What must be the minimum rating in voltamperes of the power source now?

(a) $f_0 = \dfrac{1}{2\pi\sqrt{LC}} = \dfrac{1}{2\pi\sqrt{0.3 \text{ H} \times 50 \times 10^{-6} \text{ F}}} = 41 \text{ Hz}$

At the resonant frequency, $X_L = X_C$ and $Z = R$. Hence

(b) $I = \dfrac{V}{R} = \dfrac{120 \text{ V}}{80 \ \Omega} = 1.5 \text{ A}$

(c) $\cos\phi = \dfrac{R}{Z} = \dfrac{R}{R} = 1 = 100\%$

(d)   True power $= IV \cos\phi = 1.5 \text{ A} \times 120 \text{ V} \times 1 = 180 \text{ W}$

(e)   Apparent power $= IV = 1.5 \text{ A} \times 120 \text{ V} = 180 \text{ V} \cdot \text{A}$

Problem 13.16. (a) Find the potential difference across the resistor, the inductor, and the capacitor in the circuit of Problem 13.13 when it is connected to a 120-V ac source whose frequency is equal to the circuit's resonant frequency of 41 Hz. (b) Are these values in accord with the applied potential difference of 120 V?

(a)   At the resonant frequency of $f_0 = 41$ Hz,   the inductive and capacitive reactances are respectively

$$X_L = 2\pi f_0 L = 2\pi \times 41 \text{ Hz} \times 0.3 \text{ H} = 77 \ \Omega$$

$$X_C = \frac{1}{2\pi f_0 C} = \frac{1}{2\pi \times 41 \text{ Hz} \times 50 \times 10^{-6} \text{ F}} = 77 \ \Omega$$

The various potential differences are therefore

$$V_R = IR = 1.5 \text{ A} \times 80 \ \Omega = 120 \text{ V}$$

$$V_L = IX_L = 1.5 \text{ A} \times 77 \ \Omega = 116 \text{ V}$$

$$V_C = IX_C = 1.5 \text{ A} \times 77 \ \Omega = 116 \text{ V}$$

(b)   The total potential difference across the circuit is

$$V = \sqrt{V_R^2 + (V_L - V_C)^2} = \sqrt{(120 \text{ V})^2 + (0)^2} = 120 \text{ V}$$

which is equal to the applied potential difference.

Problem 13.17. In a series ac circuit $R = 20 \ \Omega$, $X_L = 10 \ \Omega$,   and $X_C = 25 \ \Omega$ when the frequency is 400 Hz.   (a) Find the impedance of the circuit.   (b) Find the phase angle.   (c) Is the resonant frequency of the circuit greater than or less than 400 Hz?   (d) Find the resonant frequency.

(a)
$$Z = \sqrt{R^2 + (X_L - X_C)^2} = \sqrt{(20 \ \Omega)^2 + (10 \ \Omega - 25 \ \Omega)^2} = 25 \ \Omega$$

(b)
$$\tan\phi = \frac{X_L - X_C}{R} = \frac{10 \ \Omega - 25 \ \Omega}{20 \ \Omega} = -0.75$$

$$\phi = -37°$$

A negative phase angle signifies that the voltage lags behind the current.

(c)   At resonance $X_L = X_C$.   At 400 Hz, $X_L < X_C$,   so the frequency must be changed in such a way as to increase X_L and decrease X_C.   Since $X_L = 2\pi f L$ and $X_C = 1/2\pi f C$,   it is clear that increasing the frequency will have this effect.   Hence the resonant frequency must be greater than 400 Hz.

(d)   Since $X_L = 10 \ \Omega$ and $X_C = 25 \ \Omega$ when $f = 400$ Hz,

$$L = \frac{X_L}{2\pi f} = \frac{10 \ \Omega}{2\pi \times 400 \text{ Hz}} = 4 \times 10^{-3} \text{ H}$$

$$C = \frac{1}{2\pi f X_C} = \frac{1}{2\pi \times 400 \text{ Hz} \times 25 \ \Omega} = 1.6 \times 10^{-5} \text{ F}$$

Hence

$$f_0 = \frac{1}{2\pi\sqrt{LC}} = \frac{1}{2\pi\sqrt{4 \times 10^{-3} \text{ H} \times 1.6 \times 10^{-5} \text{ F}}} = 629 \text{ Hz}$$

Problem 13.18. A 5-hp electric motor is 80 percent efficient and has an inductive power factor of 75 percent.   (a) What minimum rating in kV·A must its power source have?   (b) A capacitor is connected in series with the motor to raise the power factor to 100 percent.   What minimum rating in kV·A must the power source now have?

(a)   The power required by the motor is

$$P = \frac{5 \text{ hp} \times 0.746 \text{ kW/hp}}{0.8} = 4.66 \text{ kW}$$

Since $P = IV \cos\phi$,   the power source must have the minimum rating

$$IV = \frac{P}{\cos\phi} = \frac{4.66 \text{ kW}}{0.75} = 6.21 \text{ kV·A}$$

(b)   When $\cos\phi = 1$, $IV = P = 4.66$ kV·A.

Problem 13.19. A coil connected to a 120-V, 25-Hz power line draws a current of 0.5 A and dissipates 50 W. (a) What is its power factor? (b) What capacitance should be connected in series with the coil to increase the power factor to 100 percent? (c) What would the current in the circuit be then? (d) How much power would the circuit then dissipate?

(a) Since $P = IV \cos \phi$,

$$\cos \phi = \frac{P}{IV} = \frac{50 \text{ W}}{0.5 \text{ A} \times 120 \text{ V}} = 0.833 = 83.3 \text{ percent}$$

(b) The power factor will be 100 percent at resonance, when $X_L = X_C$. The first step is to find X_L, which can be done from the formula $\tan \phi = (X_L - X_C)/R$. Here $X_C = 0$ and, since $\cos \phi = 0.833$, $\phi = 34°$ and $\tan \phi = 0.663$. Since $P = I^2 R$,

$$R = \frac{P}{I^2} = \frac{50 \text{ W}}{(0.5 \text{ A})^2} = 200 \ \Omega$$

Hence

$$X_L = R \tan \phi + X_C = 200 \ \Omega \times 0.663 + 0 = 133 \ \Omega$$

This must also be the value of X_C when $f = 25$ Hz, and so

$$C = \frac{1}{2\pi f X_C} = \frac{1}{2\pi \times 25 \text{ Hz} \times 133 \ \Omega} = 4.8 \times 10^{-5} \text{ F} = 48 \ \mu F$$

(c) $Z = R$ at resonance, so

$$I = \frac{V}{Z} = \frac{V}{R} = \frac{120 \text{ V}}{200 \ \Omega} = 0.6 \text{ A}$$

(d)

$$P = I^2 R = (0.6 \text{ A})^2 \times 200 \ \Omega = 72 \text{ W}$$

Supplementary Problems

13.20. Find the reactance of a 10-mH inductor when it is in a 500-Hz circuit.

13.21. The reactance of an inductor is 1000 Ω at 200 Hz. Find its inductance.

13.22. A current of 0.20 A flows through a 0.15-H inductor of negligible resistance that is connected to an 80-V source of alternating current. What is the frequency of the source?

13.23. Find the reactance of a 5-μF capacitor at 10 Hz and at 10 kHz.

13.24. A capacitor has a reactance of 200 Ω at 1000 Hz. What is its capacitance?

13.25. A 5-μF capacitor is connected to a 6-kHz alternating emf, and a current of 2 A flows. Find the effective magnitude of the emf.

13.26. A 5-μF capacitor draws 1 A when connected to a 60-V ac source. What is the frequency of the source?

13.27. A 25-μF capacitor is connected in series with a 50-Ω resistor and a 12-V, 60-Hz potential difference is applied. Find the current in the circuit and the power dissipated in it.

13.28. A 0.1-H, 30-Ω inductor is connected to a 50-V, 100-Hz power source. Find the current in the inductor and the power dissipated in it.

13.29. A capacitor is connected in series with an 8-Ω resistor and the combination is placed across a 24-V, 1000-Hz power source. A current of 2 A flows. Find (a) the capacitance of the capacitor, (b) the phase angle, and (c) the power dissipated by the circuit.

13.30. The current in a resistor is 1 A when it is connected to a 50-V, 100-Hz power source. (a) How much inductive reactance is required to reduce the current to 0.5 A? What value of L will accomplish this? (b) How much capacitive reactance is required to reduce the current to 0.5 A? What value of C will accomplish this? (c) What will the current be if the above inductance and capacitance are both placed in series with the resistor?

13.31. A pure resistor, a pure capacitor, and a pure inductor are connected in series across an ac power source. An ac voltmeter placed in turn across these circuit elements reads 10 V, 20 V, and 30 V. What is the potential difference of the source?

13.32. An inductive circuit with a power factor of 80 percent consumes 750 W of power. What minimum rating in voltamperes must its power source have?

13.33. A 10-μF capacitor, a 10-mH inductor, and a 10-Ω resistor are connected in series with a 45-V, 400-Hz power source. Find (a) the impedance of the circuit, (b) the current in it, (c) the power it dissipates, and (d) the minimum rating in voltamperes of the power source.

13.34. (a) Find the resonant frequency f_0 of the circuit of Problem 13.33. (b) What current will flow in the circuit if it is connected to a 45-V power source whose frequency is f_0? (c) How much power will be dissipated by the circuit? (d) What must be the minimum rating of the power source in voltamperes?

13.35. A 60-μF capacitor, a 0.3-H inductor, and a 50-Ω resistor are connected in series with a 120-V, 60-Hz power source. Find (a) the impedance of the circuit, (b) the current in it, (c) the power it dissipates, and (d) the minimum rating in voltamperes of the power source.

13.36. (a) Find the resonant frequency f_0 of the circuit of Problem 13.35. (b) What current will flow in the circuit if it is connected to a 120-V power source whose frequency is f_0? (c) How much power will be dissipated by the circuit? (d) What must be the minimum rating of the power source in voltamperes?

13.37. In a series circuit $R = 100\ \Omega$, $X_L = 120\ \Omega$, and $X_C = 60\ \Omega$ when it is connected to an 80-V ac power source. Find (a) the current in the circuit, (b) the phase angle, and (c) the current if the frequency of the power source were changed to be equal to the resonant frequency of the circuit.

13.38. A circuit that consists of a capacitor in series with a 50-Ω resistor draws 4 A from a 250-V, 200 Hz power source. (a) How much power is dissipated? (b) What is the power factor? (c) What inductance should be connected in series in the circuit to increase the power factor to 100 percent? (d) What would the current then be? (e) How much power would the circuit then dissipate?

13.39. An inductive load dissipates 75 W of power when it draws 1.0 A from a 120-V, 60-Hz power line. (a) What is the power factor? (b) What capacitance should be connected in series in the circuit to increase the power factor to 100 percent? (c) What would the current then be? (d) How much power would the circuit then dissipate?

Answers to Supplementary Problems

13.20. 31.4 Ω

13.21. 0.796 H

13.22. 424 Hz

13.23. 3183 Ω; 3.183 Ω

13.24. 0.796 μF

13.25. 10.6 V

13.26. 531 Hz

13.27. 0.102 A; 0.524 W

13.28. 0.72 A; 15.6 W

13.29. (a) 17.8 μF   (b) 48°   (c) 32 W

13.30. (a) 50 Ω, 79.6 mH   (b) 50 Ω, 31.8 μF   (c) 1 A

13.31. 14 V

13.32. 938 V · A

13.33. (a) 18 Ω   (b) 2.5 A   (c) 62.5 W   (d) 112.5 V · A

13.34. (a) 503 Hz   (b) 4.5 A   (c) 202.5 W   (d) 202.5 V · A

13.35. (a) 85 Ω   (b) 1.41 A   (c) 99.4 W   (d) 169 V · A

13.36. (a) 37.5 Hz   (b) 2.4 A   (c) 288 W   (d) 288 V · A

13.37. (a) 0.69 A   (b) 31°   (c) 0.80 A

13.38. (a) 800 W   (b) 80%   (c) 30 mH   (d) 5 A   (e) 1250 W

13.39. (a) 0.625   (b) 28.6 μF   (c) 1.6 A   (d) 192 W

Chapter 14

Parallel AC Circuits

PHASE RELATIONS

When a resistor, an inductor, and a capacitor are connected in parallel across an ac source, as in Fig. 14-1, the voltage is the same across each circuit element:

$$V = V_R = V_L = V_C$$

The total instantaneous current is the sum of the instantaneous currents in each branch, as in a parallel dc circuit, but this is not true of the total effective current I because the branch currents are not in phase. Although the current I_R in the resistor is in phase with V, the current I_C in the capacitor leads V by 90° and the current I_L in the inductor lags V by 90°. To find the total current I, the phasors that represent I_R, I_C, and I_L must be added vectorially, as in Fig. 14-1(b).

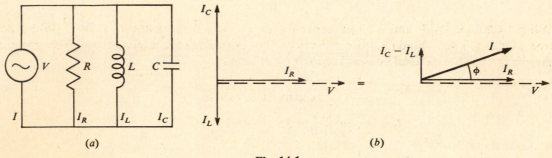

(a)   (b)

Fig. 14-1

The branch currents in the parallel circuit of Fig. 14-1 are given by

$$I_R = \frac{V}{R} \qquad I_C = \frac{V}{X_C} \qquad I_L = \frac{V}{X_L}$$

Adding these currents vectorially with the help of the Pythagorean theorem gives

$$I = \sqrt{I_R^2 + (I_C - I_L)^2}$$

The phase angle ϕ between current and voltage is specified by

$$\cos \phi = \frac{I_R}{I}$$

If I_C is greater than I_L, the current leads the voltage and the phase angle is considered positive; if I_L is greater than I_C, the current lags the voltage and the phase angle is considered negative. The power dissipated in a parallel ac circuit is given by the same formula as in a series circuit, namely

$$P = IV \cos \phi$$

Problem 14.1. A 400-Ω and a 500-Ω resistor are connected in parallel to a 240-V, 60-Hz power source, as in Fig. 14-2. Find (a) the current in each resistor, (b) the total current, (c) the total resistance of the circuit, and (d) the total power dissipated by the circuit.

(a)
$$I_1 = \frac{V}{R_1} = \frac{240 \text{ V}}{400 \text{ } \Omega} = 0.60 \text{ A} \qquad I_2 = \frac{V}{R_2} = \frac{240 \text{ V}}{500 \text{ } \Omega} = 0.48 \text{ A}$$

140

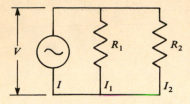

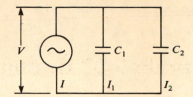

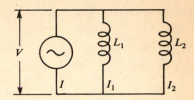

Fig. 14-2   Fig. 14-3   Fig. 14-4

(b) The currents in the resistors are in phase with each other, so that

$$I = I_1 + I_2 = 0.60 \text{ A} + 0.48 \text{ A} = 1.08 \text{ A}$$

(c)
$$R = \frac{V}{I} = \frac{240 \text{ V}}{1.08 \text{ A}} = 222 \ \Omega$$

(d) Since the current is in phase with the voltage, $\phi = 0$ and $\cos \phi = 1$. Hence

$$P = IV \cos \phi = 240 \text{ V} \times 1.08 \text{ A} \times 1 = 259 \text{ W}$$

Alternatively we could have proceeded as in Chapter 7 for a dc circuit:

$$P = I^2 R = (1.08 \text{ A})^2 \times 222 \ \Omega = 259 \text{ W}$$

Problem 14.2. A 30-μF and a 50-μF capacitor are connected in parallel to a 50-V, 400-Hz power source, as in Fig. 14-3.   Find (a) the current in each capacitor, (b) the total current, (c) the impedance of the circuit, and (d) the total power dissipated by the circuit.

(a)
$$X_{C_1} = \frac{1}{2\pi f C_1} = \frac{1}{2\pi \times 400 \text{ Hz} \times 30 \times 10^{-6} \text{ F}} = 13 \ \Omega$$

$$I_1 = \frac{V}{X_{C_1}} = \frac{50 \text{ V}}{13 \ \Omega} = 3.8 \text{ A}$$

$$X_{C_2} = \frac{1}{2\pi f C_2} = \frac{1}{2\pi \times 400 \text{ Hz} \times 50 \times 10^{-6} \text{ F}} = 8 \ \Omega$$

$$I_2 = \frac{V}{X_{C_2}} = \frac{50 \text{ V}}{8 \ \Omega} = 6.2 \text{ A}$$

(b) The currents in the capacitors are in phase with each other, so that

$$I = I_1 + I_2 = 3.8 \text{ A} + 6.2 \text{ A} = 10 \text{ A}$$

(c)
$$Z = \frac{V}{I} = \frac{50 \text{ V}}{10 \text{ A}} = 5 \ \Omega$$

(d) In a purely capacitive circuit, the current leads the voltage by 90°.   There is no component of I in phase with V, hence no power is dissipated in the circuit.   More formally, since $\phi = 90°$ and $\cos 90° = 0$,

$$W = IV \cos \phi = 50 \text{ V} \times 10 \text{ A} \times 0 = 0$$

Problem 14.3. A 5-mH and an 8-mH inductor are connected in parallel to a 50-V, 400-Hz power source, as in Fig. 14-4.   Find (a) the current in each inductor, (b) the total current, (c) the impedance of the circuit, and (d) the total power dissipated by the circuit.

(a)
$$X_{L_1} = 2\pi f L_1 = 2\pi \times 400 \text{ Hz} \times 5 \times 10^{-3} \text{ H} = 12.6$$

$$I_1 = \frac{V}{X_{L_1}} = \frac{50 \text{ V}}{12.6 \ \Omega} = 4 \text{ A}$$

$$X_{L_2} = 2\pi f L_2 = 2\pi \times 400 \text{ Hz} \times 8 \times 10^{-3} \text{ H} = 20 \ \Omega$$

$$I_2 = \frac{V}{X_{L_2}} = \frac{50 \text{ V}}{20 \ \Omega} = 2.5 \text{ A}$$

(*b*)   The currents in the inductors are in phase with each other, so that

$$I = I_1 + I_2 = 4 \text{ A} + 2.5 \text{ A} = 6.5 \text{ A}$$

(*c*)

$$Z = \frac{V}{I} = \frac{50 \text{ V}}{6.5 \text{ A}} = 7.7 \ \Omega$$

(*d*)   In a purely inductive circuit, the current lags behind the voltage by 90°. There is no component of I in phase with V, hence no power is dissipated in the circuit. As in the previous problem, since $\phi = 90°$ and $\cos 90° = 0$,

$$P = IV \cos \phi = 50 \text{ V} \times 6.5 \text{ A} \times 0 = 0$$

Problem 14.4. A 10-Ω resistor and an 8-μF capacitor are connected in parallel across a 10-V, 1000-Hz power source, as in Fig. 14-5(a). Find (a) the current in each component, (b) the total current, (c) the impedance of the circuit, and (d) the phase angle and power dissipation of the circuit.

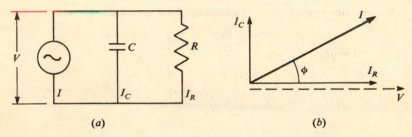

(*a*)   (*b*)

Fig. 14-5

(*a*)

$$X_C = \frac{1}{2\pi f C} = \frac{1}{2\pi \times 10^3 \text{ Hz} \times 8 \times 10^{-6} \text{ F}} = 20 \ \Omega$$

$$I_C = \frac{V}{X_C} = \frac{10 \text{ V}}{20 \ \Omega} = 0.5 \text{ A} \qquad I_R = \frac{V}{R} = \frac{10 \text{ V}}{10 \ \Omega} = 1.0 \text{ A}$$

(*b*)   The currents in the two branches cannot be added together arithmetically because they are 90° out of phase. Instead they must be combined vectorially, as in the phasor diagram of Fig. 14-5(b), to give

$$I = \sqrt{I_R^2 + I_C^2} = \sqrt{(1.0 \text{ A})^2 + (0.5 \text{ A})^2} = \sqrt{1.25 \text{ A}^2} = 1.12 \text{ A}$$

(*c*)

$$Z = \frac{V}{I} = \frac{10 \text{ V}}{1.12 \text{ A}} = 8.9 \ \Omega$$

(*d*)

$$\cos \phi = \frac{I_R}{I} = \frac{1.0 \text{ A}}{1.12 \text{ A}} = 0.893$$

$$\phi = 27°$$

The current leads the voltage in the circuit by 27°.

$$P = IV \cos \phi = 10 \text{ V} \times 1.12 \text{ A} \times 0.893 = 10 \text{ W}$$

Since all the power is dissipated in the resistor, we can also find P by considering the current in the resistor:

$$P = I_R^2 R = (1.0 \text{ A})^2 \times 10 \ \Omega = 10 \text{ W}$$

Problem 14.5. A 10-Ω resistor and a 2-mH inductor are connected in parallel across a 10-V, 1000-Hz power source, as in Fig. 14-6(a). Find (a) the current in each component, (b) the total current, (c) the impedance of the circuit, and (d) the phase angle and the total power dissipation of the circuit.

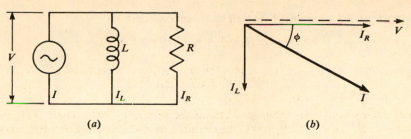

Fig. 14-6

(a)
$$X_L = 2\pi f L = 2\pi \times 10^3 \times 2 \times 10^{-3} \text{ H} = 12.6 \; \Omega$$

$$I_L = \frac{V}{X_L} = \frac{10 \text{ V}}{12.6 \; \Omega} = 0.8 \text{ A} \qquad I_R = \frac{V}{R} = \frac{10 \text{ V}}{10 \; \Omega} = 1.0 \text{ A}$$

(b) The current in the inductor lags 90° behind the current in the resistor, as shown in the phasor diagram of Fig. 14-6(b). Hence

$$I = \sqrt{I_R^2 + I_L^2} = \sqrt{(1.0 \text{ A})^2 + (0.8 \text{ A})^2} = \sqrt{1.64 \text{ A}^2} = 1.3 \text{ A}$$

(c)
$$Z = \frac{V}{I} = \frac{10 \text{ V}}{1.3 \text{ A}} = 7.7 \; \Omega$$

(d)
$$\cos \phi = \frac{I_R}{I} = \frac{1.0 \text{ A}}{1.3 \text{ A}} = 0.769$$
$$\phi = -40°$$

The current lags the voltage in the circuit by 40°.

$$P = IV \cos \phi = 10 \text{ V} \times 1.3 \text{ A} \times 0.769 = 10 \text{ W}$$

Alternatively,

$$P = I_R^2 R = (1.0 \text{ A})^2 \times 10 \; \Omega = 10 \text{ W}$$

Problem 14.6. A 10-Ω resistor, an 8-μF capacitor, and a 2-mH inductor are connected in parallel across a 10-V, 1000-Hz power source, as in Fig. 14-7(a). Find (a) the current in each component, (b) the total current in the circuit, (c) the impedance of the circuit, and (d) the phase angle and the total power dissipation of the circuit.

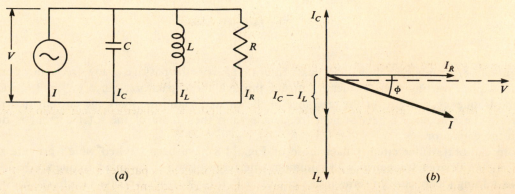

Fig. 14-7

(a) From Problems 14.4 and 14.5, $X_C = 20 \ \Omega$ and $X_L = 12.6 \ \Omega$. Hence

$$I_C = \frac{10 \text{ V}}{20 \ \Omega} = 0.5 \text{ A} \qquad I_L = \frac{10 \text{ V}}{12.6 \ \Omega} = 0.8 \text{ A} \qquad I_R = \frac{10 \text{ V}}{10 \ \Omega} = 1.0 \text{ A}$$

(b) The phasor diagram of Fig. 14-7(b) shows how the various currents are to be added. We have

$$I = \sqrt{I_A^2 + (I_C - I_L)^2} = \sqrt{(1.0 \text{ A})^2 + (0.5 \text{ A} - 0.8 \text{ A})^2}$$
$$= \sqrt{(1.0 \text{ A})^2 + (-0.3 \text{ A})^2} = \sqrt{1.00 \text{ A}^2 + 0.09 \text{ A}^2} = \sqrt{1.09 \text{ A}^2} = 1.04 \text{ A}$$

(c)
$$Z = \frac{V}{I} = \frac{10 \text{ V}}{1.04 \text{ A}} = 9.6 \ \Omega$$

(d)
$$\cos \frac{I_r}{I} = \frac{1.0 \text{ A}}{1.04 \text{ A}} = 0.962$$
$$\phi = -16°$$

The current lags behind the voltage by 16°, as in Fig. 14-7(b).

$$P = IV \cos \phi = 10 \text{ V} \times 1.04 \text{ A} \times 0.962 = 10 \text{ W}$$

We can also find P in this way:

$$P = I_R^2 R = (1.0 \text{ A})^2 \times 10 \ \Omega = 10 \text{ W}$$

FILTERS

A *filter* is a circuit designed to allow currents in a certain frequency band to pass through it relatively unimpeded while rejecting currents whose frequencies lie outside that band. Figure 14-8(a) shows a simple "high-pass" filter that permits high-frequency currents to reach the load while diverting low-frequency currents through the inductor. Since $X_L = 2\pi fL$, the higher the frequency, the greater the reactance of the inductor and the greater the proportion of the total current that gets to the load. At low frequencies the reactance is less and more of the total current goes through the inductor instead of the load.

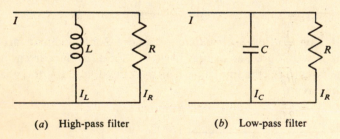

(a) High-pass filter   (b) Low-pass filter

Fig. 14-8

Figure 14-8(b) shows a simple "low-pass" filter that favors the passage of low-frequency current to the load while diverting high-frequency current through the capacitor. Since $X_C = 1/2\pi fC$, the lower the frequency, the greater the reactance of the capacitor and the more the current that reaches the load. The reactance is less at high frequencies and then more of the total current is diverted through the capacitor.

More elaborate filter circuits than those of Fig. 14-8 have been devised which provide sharper cut-offs at the desired frequency with minimum interference with the wanted frequency band. For example, adding a capacitor in series in the circuit of Fig. 14-8(a) will increase the effectiveness of the high-pass filter because the reactance of the capacitor decreases with increasing frequency. At low frequencies, X_C is large so the current that gets through is small, and since X_L is small at such frequencies, most of this current passes through the inductor to leave very little for the load R. At high frequencies, X_C is small, allowing a large current in the entire circuit, and X_L is large, so nearly all the current goes through the load. Similar reasoning shows that adding an inductor in series in the circuit of Fig. 14-8(b) will improve its low-pass ability.

Problem 14.7. Design a simple high-pass filter like that shown in Fig. 14-8(a) for a 50-Ω load that discriminates against frequencies lower than 20 kHz.

If $X_L = R$ at $f = 20$ kHz $= 20 \times 10^3$ Hz, then for the same applied voltage, I_L will be greater than I_R for frequencies less than 20 kHz, and I_L will be less than I_R for frequencies greater than 20 kHz. Hence the required inductance can be found as follows:

$$X_L = R$$
$$2\pi f L = R$$
$$L = \frac{R}{2\pi f} = \frac{50\ \Omega}{2\pi \times 20 \times 10^3\ \text{Hz}} = 4 \times 10^{-4}\text{H} = 0.4\ \text{mH}$$

Problem 14.8. Find the percentage of the total current that passes through the load in the circuit of Problem 14.7 when the frequency of the applied voltage is (a) 2 kHz and (b) 200 kHz.

(a) For convenience we can assume an applied voltage of $V = 100$ V. (We could equally well use any other voltage, or just call the applied voltage V.) We first find I_L and I_R:

$$X_L = 2\pi f L = 2\pi \times 2 \times 10^3\ \text{Hz} \times 4 \times 10^{-4}\ \text{H} = 5\ \Omega$$

$$I_L = \frac{V}{X_L} = \frac{100\ \text{V}}{5\ \Omega} = 20\ \text{A} \qquad I_R = \frac{V}{R} = \frac{100\ \text{V}}{50\ \Omega} = 2\ \text{A}$$

The total current in the circuit is

$$I = \sqrt{I_R^2 + I_L^2} = \sqrt{(2\ \text{A})^2 + (20\ \text{A})^2} = \sqrt{404\ \text{A}^2} = 20.1\ \text{A}$$

and the percentage of I that passes through the load is therefore

$$\frac{I_R}{I} = \frac{2\ \text{A}}{20.1\ \text{A}} = 0.0995 = 9.95\ \text{percent}$$

(b)
$$X_L = 2\pi f L = 2\pi \times 200 \times 10^3\ \text{Hz} \times 4 \times 10^{-4}\ \text{H} = 503\ \Omega$$

$$I_L = \frac{V}{X_L} = \frac{100\ \text{V}}{503\ \Omega} = 0.2\ \text{A} \qquad I_R = \frac{V}{R} = \frac{100\ \text{V}}{50\ \Omega} = 2\ \text{A}$$

$$I = \sqrt{I_R^2 + I_L^2} = \sqrt{(2\ \text{A})^2 + (0.2\ \text{A})^2} = \sqrt{4.04\ \text{A}^2} = 2.01\ \text{A}$$

$$\frac{I_R}{I} = \frac{2\ \text{A}}{2.01\ \text{A}} = 0.995 = 99.5\ \text{percent}$$

Problem 14.9. Design a simple low-pass filter like that in Fig. 14-8(b) that discriminates against frequencies higher than 20 kHz.

If $X_C = R$ for $f = 20$ kHz $= 20 \times 10^3$ Hz, then for the same applied voltage, I_C will be greater than I_R for frequencies greater than 20 kHz, and I_C will be less than I_R for frequencies lower than 20 kHz. Hence the required capacitance can be found as follows:

$$X_C = R$$
$$\frac{1}{2\pi f C} = R$$
$$1 = 2\pi f C R$$
$$C = \frac{1}{2\pi f R} = \frac{1}{2\pi \times 20 \times 10^3 \times 50\ \Omega} = 1.6 \times 10^{-7}\ \text{F} = 0.16 \times 10^{-6}\ \text{F} = 0.16\ \mu\text{F}$$

Problem 14.10. Find the percentage of the total current that passes through the load in the circuit of Problem 14.9 when the frequency of the applied voltage is (*a*) 2 kHz and (*b*) 200 kHz.

(*a*) As in Problem 14.8, we assume an applied voltage of 100 V and begin by finding the currents I_C and I_R.

$$X_C = \frac{1}{2\pi f C} = \frac{1}{2\pi \times 2 \times 10^3 \text{ Hz} \times 1.6 \times 10^{-7} \text{ F}} = 497 \ \Omega$$

$$I_C = \frac{V}{X_C} = \frac{100 \text{ V}}{497 \ \Omega} = 0.2 \text{ A} \qquad I_R = \frac{V}{R} = \frac{100 \text{ V}}{50 \ \Omega} = 2 \text{ A}$$

The total current in the circuit is

$$I = \sqrt{I_R^2 + I_C^2} = \sqrt{(2 \text{ A})^2 + (0.2 \text{ A})^2} = \sqrt{4.04 \text{ A}^2} = 2.01 \text{ A}$$

and the percentage that passes through the load is therefore

$$\frac{I_R}{I} = \frac{2 \text{ A}}{2.01 \text{ A}} = 0.995 = 99.5 \text{ percent}$$

(*b*)
$$X_C = \frac{1}{2\pi f C} = \frac{1}{2\pi \times 200 \times 10^3 \text{ Hz} \times 1.6 \times 10^{-7} \text{ F}} = 5 \ \Omega$$

$$I_C = \frac{V}{X_C} = \frac{100 \text{ V}}{5 \ \Omega} = 20 \text{ A} \qquad I_R = \frac{V}{R} = \frac{100 \text{ V}}{50 \ \Omega} = 2 \text{ A}$$

$$I = \sqrt{I_R^2 + I_C^2} = \sqrt{(2 \text{ A})^2 + (20 \text{ A})^2} = \sqrt{404 \text{ A}^2} = 20.1 \text{ A}$$

$$\frac{I_R}{I} = \frac{2 \text{ A}}{20.1 \text{ A}} = 0.0995 = 9.95 \text{ percent}$$

RESONANCE

Figure 14-9 shows an inductor and a capacitor connected in parallel to a power source. The currents in the inductor and capacitor are 180° apart in phase, as the phasor diagram shows, so the total current I in the circuit is the *difference* between the currents in L and C:

$$I = I_C - I_L$$

The current that circulates between the inductor and the capacitor without contributing to I is called the *tank current* and may be greater than I.

In the event that $X_C = X_L$, the currents I_C and I_L are also equal. Since I_C and I_L are 180° out of phase, the total current $I = 0$: the currents in the inductor and capacitor cancel each other out. This situation is called *resonance*.

In a series *RLC* circuit, as discussed in Chapter 13, the impedance is a minimum of $Z = R$ when $X_C = X_L$, a situation also called resonance. The frequency for which $X_C = X_L$ is

$$f_0 = \frac{1}{2\pi\sqrt{LC}}$$

and was called the resonant frequency.

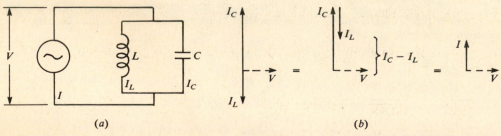

(*a*)   (*b*)

Fig. 14-9

In a parallel RLC circuit, resonance again corresponds to $X_C = X_L$, but here the impedance is a *maximum* at f_0. At f_0, the currents in the inductor and capacitor are equal in magnitude but 180° out of phase, so no current passes through the combination (Fig. 14-10). Thus $I = I_R$ and $Z = R$. At frequencies higher and lower than f_0, I_C is not equal to I_L and some current can pass through the inductor-capacitor part of the circuit, which reduces the impedance Z to less than R. Thus a series circuit can be used as a selector to favor a particular frequency, and a parallel circuit with the same L and C can be used as a selector to discriminate against the same frequency.

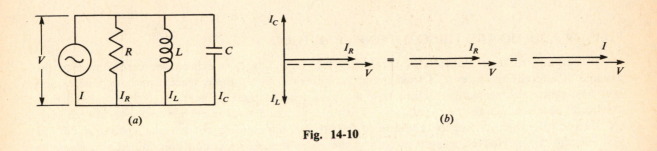

(a)   (b)

Fig. 14-10

Problem 14.11. The reactance of a coil and a capacitor connected in parallel and supplied by a 15-V, 1000-Hz power source are respectively $X_L = 20\ \Omega$ and $X_C = 30\ \Omega$ (see Fig. 14-9). Find (a) the currents in each component, (b) the total current, (c) the impedance of the circuit, and (d) the phase angle and total power dissipated in the circuit.

(a)
$$I_L = \frac{V}{X_L} = \frac{15\ \text{V}}{20\ \Omega} = 0.75\ \text{A} \qquad I_C = \frac{V}{X_C} = \frac{15\ \text{V}}{30\ \Omega} = 0.5\ \text{A}$$

(b)
$$I = I_C - I_L = 0.5\ \text{A} - 0.75\ \text{A} = -0.25\ \text{A}$$

The minus sign means that the total current lags 90° behind the voltage [the opposite of the situation shown in Fig. 14-9(b)], and can be disregarded in the other calculations.

(c)
$$Z = \frac{V}{I} = \frac{15\ \text{V}}{0.25\ \text{A}} = 60\ \Omega$$

The impedance is not only greater than X_L or X_C but is greater than their arithmetical sum.

(d) Because the phase angle here is 90°, $\cos \phi = \cos 90° = 0$ and the power drawn by the circuit is

$$P = IV \cos \phi = 0$$

This conclusion follows from the absence of resistance in the circuit.

Problem 14.12. The antenna circuit of a radio receiver contains a "wave trap" that consists of a 100-pF capacitor and a 180-μH coil connected in parallel. What is the resonant frequency of the trap?

Since $L = 180\ \mu\text{H} = 180 \times 10^{-6}\ \text{H} = 1.8 \times 10^{-4}\ \text{H}$ and $C = 100\ \text{pF} = 100 \times 10^{-12}\ \text{F} = 10^{-10}\ \text{F}$, the resonant frequency is

$$f_0 = \frac{1}{2\pi\sqrt{LC}} = \frac{1}{2\pi\sqrt{1.8 \times 10^{-4}\ \text{H} \times 10^{-10}\ \text{F}}} = \frac{1}{2\pi\sqrt{1.8}\ \sqrt{10^{-4} \times 10^{-10}}}\ \text{Hz} = \frac{1}{2\pi\sqrt{1.8}\ \sqrt{10^{-14}}}\ \text{Hz}$$

$$= \frac{1}{2\pi \times 1.34 \times 10^{-7}}\ \text{Hz} = 0.12 \times 10^7\ \text{Hz} = 1.2 \times 10^6\ \text{Hz} = 1.2\ \text{MHz}$$

Problem 14.13. What should the capacitance of the capacitor of Problem 14.12 be changed to in order to have the resonant frequency of the trap be 1.5 MHz?

$$f_0 = \frac{1}{2\pi\sqrt{LC}}$$

$$f_0^2 = \frac{1}{4\pi^2 LC}$$

$$C = \frac{1}{4\pi^2 L f_0^2} = \frac{1}{4\pi^2 \times 1.8 \times 10^{-4} \text{ H} \times (1.5 \times 10^6 \text{ Hz})^2} = 62.5 \times 10^{-12} \text{ F} = 62.5 \text{ pF}$$

PHASOR ADDITION BY THE COMPONENT METHOD

In addition to their reactances, actual inductors and capacitors exhibit some resistance to the passage of alternating current. Other circuit components, for example, electric motors, also exhibit both resistance and reactance. Until now we have ignored these resistances and considered only idealized "pure" inductors and capacitors. The phase lags and leads in such inductors and capacitors are always 90°, which makes the addition of the phasors representing current and voltage in circuits that contain them relatively easy. In reality, the phase angle in a circuit component can never be exactly 90° because of the presence of resistance, and we must know how to add phasors at other angles.

The most straightforward way to add phasors at arbitrary angles is the component method, which was described in Chapter 11 as applied to vectors. The procedure is exactly the same here, however:

1. Resolve the initial phasors into their components in the y (vertical) and x (horizontal) directions, as shown in Fig. 14-11(a) and (b) for the current phasors I_A and I_B in a parallel circuit. The x direction is the same as that of the total voltage V in the circuit. The components are

$$I_{Ax} = I_A \cos\phi_A \qquad I_{Bx} = I_B \cos\phi_B$$

$$I_{Ay} = I_A \sin\phi_A \qquad I_{By} = I_B \sin\phi_B$$

2. Add the components in the x direction together to give I_x (which is equal to I_R) and add the components in the y direction to give I_y, as in Fig. 14-11(c) and (d):

$$I_x = I_{Ax} + I_{Bx} = I_R \qquad I_y = I_{Ay} + I_{By}$$

3. Calculate the magnitude of the resultant current I by the Pythagorean theorem:

$$I = \sqrt{I_x^2 + I_y^2}$$

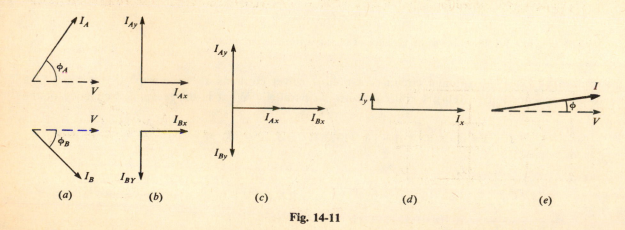

(a) (b) (c) (d) (e)

Fig. 14-11

The phase angle ϕ of Fig. 14-11(e) can be found from the formula

$$\cos\phi = \frac{I_x}{I} = \frac{I_R}{I}$$

Problem 14.14. A current of 8 A in one branch of an ac circuit leads the voltage by 60°, as in Fig. 14-12. Find the component of the current in phase with the voltage and the "reactive" component which is 90° out of phase with the voltage.

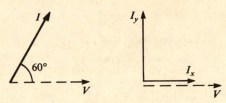

Fig. 14-12

The component of I in phase with the voltage is

$$I_x = I \cos\phi = 8 \text{ A} \times \cos 60° = 8 \text{ A} \times 0.5 = 4 \text{ A}$$

The component of I that is 90° out of phase with the voltage is

$$I_y = I \sin\phi = 8 \text{ A} \times \sin 60° = 8 \text{ A} \times 0.866 = 6.9 \text{ A}$$

The arithmetic sum of I_x and I_y is 10.9 A, which is more than the actual total current of 8 A, but this sum means nothing because I_x and I_y are out-of-phase with each other. The proper way to add I_x and I_y is

$$I = \sqrt{I_x^2 + I_y^2} = \sqrt{(4 \text{ A})^2 + (6.9 \text{ A})^2} = \sqrt{16 \text{ A}^2 + 48 \text{ A}^2} = \sqrt{64 \text{ A}^2} = 8 \text{ A}$$

which is the actual current.

Problem 14.15. In the parallel ac circuit of Fig. 14-13, a 12-A current in branch A leads the voltage of 50 V by 80° and a 10-A current in branch B lags the voltage by 50°. Find (a) the total current in the circuit, (b) the phase angle between current and voltage, and (c) the power dissipated by the circuit.

(a)
$$I_{Ax} = I_A \cos\phi_A = 12 \text{ A} \times \cos 80° = 12 \text{ A} \times 0.174 = 2.09 \text{ A}$$
$$I_{Ay} = I_A \sin\phi_A = 12 \text{ A} \times \sin 80° = 12 \text{ A} \times 0.985 = 11.82 \text{ A}$$
$$I_{Bx} = I_B \cos\phi_B = 10 \text{ A} \times \cos 50° = 10 \text{ A} \times 0.643 = 6.43 \text{ A}$$
$$I_{By} = I_B \sin\phi_B = -10 \text{ A} \times \sin 50° = -10 \text{ A} \times 0.766 = -7.66 \text{ A}$$

The minus sign is needed for I_{By} because it is in the $-y$ (downward) direction, corresponding to the current

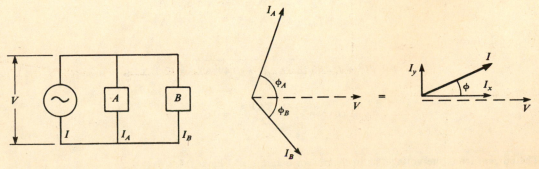

Fig. 14-13

lagging the voltage.   The total currents in the x and y directions are respectively

$$I_x = I_{Ax} + I_{Bx} = 2.09 \text{ A} + 6.43 \text{ A} = 8.52 \text{ A}$$
$$I_y = I_{Ay} + I_{By} = 11.82 \text{ A} - 7.66 \text{ A} = 4.16 \text{ A}$$

The total current is

$$I = \sqrt{I_x^2 + I_y^2} = \sqrt{(8.52 \text{ A})^2 + (4.16 \text{ A})^2} = 9.48 \text{ A}$$

(b)

$$\cos \phi = \frac{I_x}{I} = \frac{8.52 \text{ A}}{9.48 \text{ A}} = 0.899$$

$$\phi = 26°$$

The current leads the voltage, as in the phasor diagram of Fig. 14-13.

(c)

$$P = IV \cos \phi = 9.48 \text{ A} \times 50 \text{ V} \times 0.899 = 426 \text{ W}$$

Problem 14.16.   In the parallel ac circuit of Fig. 14-14, a 5-A current in branch A leads the voltage of 120 V by 58° and an 8-A current in branch B leads the voltage by 41°.   Find (a) the total current in the circuit, (b) the phase angle between current and voltage, and (c) the power dissipated by the circuit.

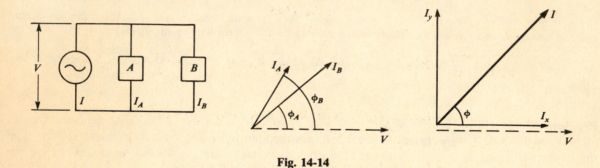

Fig. 14-14

(a)

$$I_{Ax} = I_A \cos \phi_A = 5 \text{ A} \times \cos 58° = 5 \text{ A} \times 0.530 = 2.65 \text{ A}$$
$$I_{Ay} = I_A \sin \phi_A = 5 \text{ A} \times \sin 58° = 5 \text{ A} \times 0.848 = 4.24 \text{ A}$$
$$I_{Bx} = I_B \cos \phi_B = 8 \text{ A} \times \cos 41° = 8 \text{ A} \times 0.755 = 6.04 \text{ A}$$
$$I_{By} = I_B \sin \phi_B = 8 \text{ A} \times \sin 41° = 8 \text{ A} \times 0.656 = 5.25 \text{ A}$$

The total currents in the x and y directions are respectively

$$I_x = I_{Ax} + I_{Bx} = 2.65 \text{ A} + 6.04 \text{ A} = 8.69 \text{ A}$$
$$I_y = I_{Ay} + I_{By} = 4.24 \text{ A} + 5.25 \text{ A} = 9.49 \text{ A}$$

The total current is

$$I = \sqrt{I_x^2 + I_y^2} = \sqrt{(8.69 \text{ A})^2 + (9.49 \text{ A})^2} = 12.9 \text{ A}$$

(b)

$$\cos \phi = \frac{I_x}{I} = \frac{8.69 \text{ A}}{12.9 \text{ A}} = 0.674$$

$$\phi = 48°$$

The current leads the voltage, as in the phasor diagram of Fig. 14-14.

(c)

$$P = IV \cos \phi = 12.9 \text{ A} \times 120 \text{ V} \times 0.674 = 1043 \text{ W}$$

Problem 14.17.   In a synchronous ac electric motor, the current leads the applied voltage, and in an induction ac electric motor, the current lags the applied voltage.   Figure 14-15 shows a circuit in which a synchronous motor A is connected in parallel with an induction motor B and a purely resistive load C.   Find (a) the current in each branch of the circuit, (b) the total current, (c) the phase angle between the total current and the applied voltage, and (d) the power drawn by the circuit.

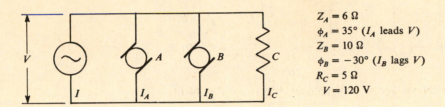

$$Z_A = 6\ \Omega$$
$$\phi_A = 35° \ (I_A \text{ leads } V)$$
$$Z_B = 10\ \Omega$$
$$\phi_B = -30° \ (I_B \text{ lags } V)$$
$$R_C = 5\ \Omega$$
$$V = 120\ \text{V}$$

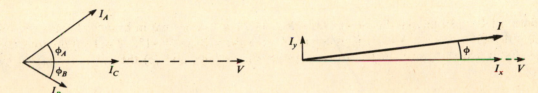

Fig. 14-15

(a) $I_A = \dfrac{V}{Z_A} = \dfrac{120\ \text{V}}{6\ \Omega} = 20\ \text{A}$ $I_B = \dfrac{V}{Z_B} = \dfrac{120\ \text{V}}{10\ \Omega} = 12\ \text{A}$ $I_C = \dfrac{V}{R_C} = \dfrac{120\ \text{V}}{5\ \Omega} = 24\ \text{A}$

(b)

$$I_{Ax} = I_A \cos \phi_A = 20\ \text{A} \times \cos 35° = 20\ \text{A} \times 0.819 = 16.4\ \text{A}$$

$$I_{Ay} = I_A \sin \phi_A = 20\ \text{A} \times \sin 35° = 20\ \text{A} \times 0.574 = 11.5\ \text{A}$$

$$I_{Bx} = I_B \cos \phi_B = 12\ \text{A} \times \cos 30° = 12\ \text{A} \times 0.866 = 10.4\ \text{A}$$

$$I_{By} = I_B \sin \phi_B = -12\ \text{A} \times \sin 30° = -12\ \text{A} \times 0.500 = -6.0\ \text{A}$$

A minus sign is used for I_{By} because I_B lags V.   The current in the resistor is in phase with the voltage, hence $\phi_C = 0$ and

$$I_{Cx} = I_C \cos \phi_C = 24\ \text{A} \times \cos 0 = 24\ \text{A} \times 1.000 = 24\ \text{A}$$
$$I_{Cy} = I_C \sin \phi_C = 24\ \text{V} \times \sin 0 = 24\ \text{A} \times 0 = 0$$

The total currents in the x and y directions are respectively

$$I_x = I_{Ax} + I_{Bx} + I_{Cx} = 16.4\ \text{A} + 10.4\ \text{A} + 24\ \text{A} = 50.8\ \text{A}$$
$$I_y = I_{Ay} + I_{By} + I_{Cy} = 11.5\ \text{A} - 6.0\ \text{A} + 0 = 5.5\ \text{A}$$

$$I = \sqrt{I_x^2 + I_y^2} = \sqrt{(50.8\ \text{A})^2 + (5.5\ \text{A})^2} = 51.1\ \text{A}$$

(c) $\cos \phi = \dfrac{I_x}{I} = \dfrac{50.8\ \text{A}}{51.1\ \text{A}} = 0.994$

$$\phi = 6°$$

The current leads the voltage.

(d) $P = IV \cos \phi = 51.1\ \text{A} \times 120\ \text{V} \times 0.994 = 6095\ \text{W}$

Problem 14.18. In the parallel ac circuit of Fig. 14-16(a), find (a) the current in each branch, (b) the total current, (c) the phase angle between the total current and the applied voltage, and (d) the power drawn by the circuit.

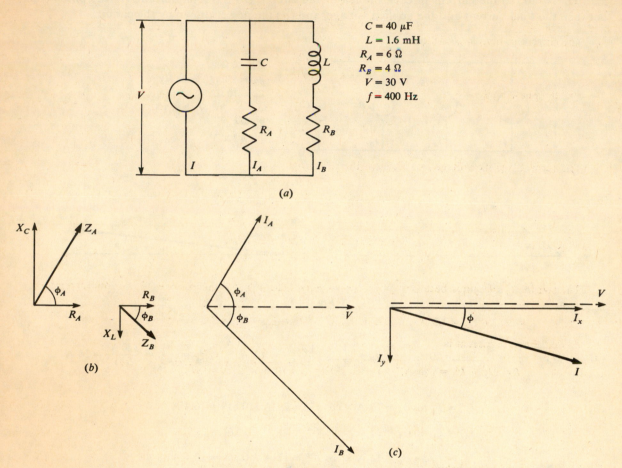

$$C = 40 \ \mu\text{F}$$
$$L = 1.6 \ \text{mH}$$
$$R_A = 6 \ \Omega$$
$$R_B = 4 \ \Omega$$
$$V = 30 \ \text{V}$$
$$f = 400 \ \text{Hz}$$

(a)

(b)

(c)

Fig. 14-16

(a) In branch A, since $C = 40 \ \mu\text{F} = 40 \times 10^{-6} \ \text{F} = 4 \times 10^{-5} \ \text{F}$ and $f = 400 \ \text{Hz} = 4 \times 10^2 \ \text{Hz}$,

$$X_C = \frac{1}{2\pi fC} = \frac{1}{2\pi \times 4 \times 10^2 \ \text{Hz} \times 4 \times 10^{-5} \ \text{F}} = \frac{1}{(2\pi \times 4 \times 4)(10^2 \times 10^{-5})} \ \Omega$$

$$= \frac{1}{2\pi \times 4 \times 4} \times \frac{1}{10^{-3}} \ \Omega = 0.0099 \times 10^3 \ \Omega = 9.9 \ \Omega$$

Branch A is a series circuit consisting of C and R_A, so its impedance is

$$Z_A = \sqrt{R_A^2 + X_C^2} = \sqrt{(6 \ \Omega)^2 + (9.9 \ \Omega)^2} = 11.6 \ \Omega$$

The current in branch A is therefore

$$I_A = \frac{V}{Z_A} = \frac{30 \ \text{V}}{11.6 \ \Omega} = 2.6 \ \text{A}$$

and the phase angle is

$$\cos \phi_A = \frac{R_A}{Z_A} = \frac{6 \ \Omega}{11.6 \ \Omega} = 0.517$$

$$\phi_A = 59°$$

The current leads the voltage by 59°, as in Fig. 14-16(b).

In branch B, since $L = 1.6$ mH $= 1.6 \times 10^{-3}$ H and $f = 400$ Hz $= 4 \times 10^2$ Hz,

$$X_L = 2\pi f L = 2\pi \times 4 \times 10^2 \text{ Hz} \times 1.6 \times 10^{-3} \text{ H} = 4 \ \Omega$$

Branch B is a series circuit consisting of L and R_B, so its impedance is

$$Z_B = \sqrt{R_B^2 + X_L^2} = \sqrt{(4 \ \Omega)^2 + (4 \ \Omega)^2} = 5.7 \ \Omega$$

The current in branch B is therefore

$$I_B = \frac{V}{Z_B} = \frac{30 \text{ V}}{5.7 \ \Omega} = 5.3 \text{ A}$$

and the phase angle is

$$\cos \phi_B = \frac{R_B}{Z} = \frac{4 \ \Omega}{5.7 \ \Omega} = 0.702$$

$$\phi_B = -45°$$

The current lags the voltage by 45°, as in Fig. 14-16(b).

(b)
$$I_{Ax} = I_A \cos \phi_A = 2.6 \text{ A} \times \cos 50° = 1.3 \text{ A}$$
$$I_{Ay} = I_A \sin \phi_A = 2.6 \text{ A} \times \sin 59° = 2.2 \text{ A}$$
$$I_{Bx} = I_B \cos \phi_B = 5.3 \text{ A} \times \cos 45° = 3.7 \text{ A}$$
$$I_{By} = I_B \sin \phi_B = -5.3 \text{ A} \times \sin 45° = -3.7 \text{ A}$$

The sign of I_{By} is minus because I_B lags the voltage. The x and y components of the total current I are respectively

$$I_x = I_{Ax} + I_{Bx} = 1.3 \text{ A} + 3.7 \text{ A} = 5 \text{ A} \qquad I_y = I_{Ay} + I_{By} = 2.2 \text{ A} - 3.7 \text{ A} = -1.5 \text{ A}$$

so that the total current is

$$I = \sqrt{I_x^2 + I_y^2} = \sqrt{(5 \text{ A})^2 + (-1.5 \text{ A})^2} = \sqrt{25 \text{ A}^2 + 2.25 \text{ A}^2} = \sqrt{27.25 \text{ A}^2} = 5.2 \text{ A}$$

(c)
$$\cos \phi = \frac{I_x}{I} = \frac{5 \text{ A}}{5.2 \text{ A}} = 0.962$$

$$\phi = -16°$$

Since the current lags behind the voltage, the phase angle is considered negative.

(d)
$$P = IV \cos \phi = 5.2 \text{ A} \times 30 \text{ V} \times 0.962 = 150 \text{ W}$$

Supplementary Problems

In Problems 14.19–14.30, find (a) the current in each component, (b) the total current in the circuit, (c) the impedance of the circuit, (d) the phase angle between current and voltage, and (e) the total power dissipated by the circuit.

14.19. Two 10-μF capacitors connected in parallel to a 12-V, 60-Hz source

14.20. A 2-μF, a 3-μF, and a 4-μF capacitor connected to a 40-V, 200-Hz source

14.21. A 10-mH and a 30-mH coil connected in parallel to a 24-V, 500-Hz source

14.22. A 100-Ω resistor and a 5-μF capacitor connected in parallel to a 24-V, 500-Hz source

14.23. A 10-Ω resistor, a 20-Ω resistor, and a 10-μF capacitor connected in parallel to a 120-V, 1000-Hz source

14.24. A 20-Ω resistor and a 5-mH coil connected in parallel to a 50-V, 1000-Hz source

14.25. A 25-Ω resistor, a 40-mH coil, and a 60-mH coil connected in parallel to a 40-V, 60-Hz source

14.26. A 10-μF capacitor and a 50-mH coil connected in parallel to a 12-V, 200-Hz source

14.27. A 5.96-μF capacitor and a 4.25-mH coil connected in parallel to a 24-V, 1000-Hz source

14.28. A 25-Ω resistor, a 40-μF capacitor, and a 40-mH coil connected in parallel to a 24-V, 100-Hz source

14.29. A 25-Ω resistor, a 40-μF capacitor, and a 40-mH coil connected in parallel to a 24-V, 200-Hz source

14.30. A 15-Ω resistor, a 5-μF capacitor, and a 2-mH coil connected in parallel to a 5-V, 2500 Hz source

14.31. What is the inductance required in a high-pass filter like that of Fig. 14-8(a) for a 600-Ω load that is to discriminate against frequencies lower than 5 kHz?

14.32. What is the capacitance required in a low-pass filter like that of Fig. 14-8(b) for a 600-Ω load that is to discriminate against frequencies higher than 5 kHz?

14.33. A high-pass filter like that of Fig. 14-8(a) has an inductance of 5 mH. When the filter is used with a load of 50 Ω, what is the frequency at which half the total current goes through the load?

14.34. A low-pass filter like that of Fig. 14-8(b) has a capacitance of 2 μF. When the filter is used with a load of 50 Ω, what is the frequency at which half the total current goes through the load?

14.35. A 0.5-μF capacitor is in parallel with a 3-mH coil. What is the resonant frequency of the circuit?

14.36. What inductance is needed in parallel with a 600-pF capacitor to make a wave trap with a resonant frequency of 300 kHz?

14.37. What capacitance is needed in parallel with a 10-mH coil to make a wave trap with a resonant frequency of 5 kHz?

14.38. A current of 60 mA in one branch of an ac circuit leads the voltage by 35°. Find the component of the current in phase with the voltage and the component that is 90° ahead of the voltage.

14.39. A current of 12 A in one branch of an ac circuit lags the voltage by 50°. Find the component of the current in phase with the voltage and the component that is 90° behind the voltage.

In Problems 14.40–14.45, find (a) the total current in each circuit, (b) the phase angle between current and voltage, and (c) the power dissipated by the circuit.

14.40. In one branch of a parallel ac circuit, a 5-A current leads the voltage of 12 V by 50°, and in the other branch a 3-A current lags the voltage by 30°.

14.41. In one branch of a parallel ac circuit, a 3-A current leads the voltage of 12 V by 30°, and in the other branch a 5-A current lags the voltage by 50°.

14.42. In one branch of a parallel ac circuit, a 3-A current leads the voltage of 12 V by 30°, and in the other branch a 5-A current leads the voltage by 50°.

14.43. In one branch of a parallel ac circuit, a 3-A current lags the voltage of 12 V by 30°, and in the other branch a 5-A current lags the voltage by 50°.

14.44. One branch of a parallel circuit consists of a 5-μF capacitor in series with a 2-Ω resistor, and the other branch consists of a 0.6-mH coil in series with a 3-Ω resistor. The circuit is connected to a 20-V, 4000-Hz source.

14.45. One branch of a parallel circuit consists of a 20-μF capacitor in series with a 30-Ω resistor, and the other branch consists of a 50-mH coil in series with a 20-Ω resistor. The circuit is connected to a 100-V, 100-Hz source.

Answers to Supplementary Problems

14.19. (a) 0.045 A; 0.045 A   (b) 0.090 A   (c) 133 Ω (d) I leads V by 90°   (e) 0

14.20. (a) 0.10 A; 0.15 A; 0.20 A   (b) 0.45 A   (c) 89 Ω (d) I leads V by 90°   (e) 0

14.21. (a) 0.764 A; 0.255 A   (b) 1.019 A   (c) 23.6 Ω (d) I lags V by 90°   (e) 0

14.22. (a) 0.240 A, 0.377 A   (b) 0.447 A   (c) 53.7 Ω (d) I leads V by 57°   (e) 5.76 W

14.23. (a) 12 A; 6 A; 7.54 A   (b) 19.52 A   (c) 6.15 Ω (d) I leads V by 23°   (e) 2.16 kW

14.24. (a) 2.50 A; 1.59 A   (b) 2.96 A   (c) 16.9 Ω (d) I lags V by 32°   (e) 125 W

14.25. (a) 1.6 A; 2.65 A; 1.77 A   (b) 4.70 A   (c) 8.51 Ω (d) I lags V by 70°   (e) 64 W

14.26. (a) 0.151 A; 0.191 A   (b) 0.040 A   (c) 300 Ω (d) I lags V by 90°   (e) 0

14.27. (a) 0.899 A; 0.899 A   (b) 0   (c) Infinite impedance   (d) No current, so no phase angle   (e) 0

14.28. (a) 0.960 A; 0.603 A; 0.955 A   (b) 1.022 A   (c) 23.48 Ω (d) I lags V by 20°   (e) 23 W

14.29. (a) 0.960 A; 1.206 A; 0.477 A   (b) 1.205 A   (c) 19.91 Ω (d) I leads V by 37°   (e) 23 W

14.30. (a) 0.333 A; 0.393 A; 0.159 A   (b) 0.407 A   (c) 12.3 Ω (d) I leads V by 35°   (e) 1.66 W

14.31. 19 mH   **14.36.** 0.469 mH

14.32. 0.053 μF   **14.37.** 0.564 μF

14.33. 1592 Hz   **14.38.** 49 mA; 34 mA

14.34. 1592 Hz   **14.39.** 7.71 A; 9.19 A

14.35. 4109 Hz

14.40.   (*a*) 6.26 A   (*b*) *I* leads *V* by 22°   (*c*) 70 W

14.41.   (*a*) 6.26 A   (*b*) *I* lags *V* by 22°   (*c*) 70 W

14.42.   (*a*) 7.89 A   (*b*) *I* leads *V* by 43°   (*c*) 70 W

14.43.   (*a*) 7.89 A   (*b*) *I* lags *V* by 43°   (*c*) 70 W

14.44.   (*a*) 1.38 A   (*b*) *I* leads *V* by 52°   (*c*) 17 W

14.45.   (*a*) 2.19 A   (*b*) *I* lags *V* by 32°   (*c*) 186 W

Appendix A

Conversion Factors

Time

1 day = 1.44×10^3 min = 8.64×10^4 s

1 year = 8.76×10^3 h = 5.26×10^5 min = 3.15×10^7 s

Length

1 meter (m) = 100 cm = 39.4 in = 3.28 ft

1 centimeter (cm) = 10 millimeters (mm) = 0.394 in

1 kilometer (km) = 10^3 m = 0.621 mi

1 foot (ft) = 12 in = 0.305 m = 30.5 cm

1 inch (in) = 0.0833 ft = 2.54 cm = 0.0254 m

1 mile (mi) = 5280 ft = 1.61 km

Area

1 m^2 = 10^4 cm^2 = 1.55×10^3 in^2 = 10.76 ft^2

1 cm^2 = 10^{-4} m^2 = 0.155 in^2

1 ft^2 = 144 in^2 = 9.29×10^{-2} m^2 = 929 cm^2

Volume

1 m^3 = 10^3 liters = 10^6 cm^3 = 35.3 ft^3 = 6.10×10^4 in^3

1 ft^3 = 1728 in^3 = 2.83×10^{-2} m^3 = 28.3 liters

Velocity

1 m/s = 3.28 ft/s = 2.24 mi/h = 3.60 km/h

1 ft/s = 0.305 m/s = 0.682 mi/h = 1.10 km/h

(*Note:* It is often convenient to remember that 88 ft/s = 60 mi/h.)

1 km/h = 0.278 m/s = 0.913 ft/s = 0.621 mi/h

1 mi/h = 1.47 ft/s = 0.447 m/s = 1.61 km/h

Mass

1 kilogram (kg) = 10^3 grams (g) = 0.0685 slug

(*Note:* 1 kg corresponds to 2.21 lb in the sense that the *weight* of 1 kg at the earth's surface is 2.21 lb.)

1 slug = 14.6 kg

(*Note:* 1 slug corresponds to 32.2 lb in the sense that the *weight* of 1 slug at the earth's surface is 32.2 lb.)

Force

1 newton (N) = 0.225 lb = 3.60 oz

1 pound (lb) = 16 ounces (oz) = 4.45 N

(*Note:* 1 lb corresponds to 0.454 kg = 454 g in the sense that the *mass* of something that weighs 1 lb at the earth's surface is 0.454 kg.)

Pressure

1 pascal (Pa) = 1 N/m^2 = 2.09×10^{-2} lb/ft^2 = 1.45×10^{-4} lb/in^2

1 lb/in^2 = 144 lb/ft^2 = 6.90×10^3 N/m^2

1 atm = 1.013×10^5 N/m^2 = 14.7 lb/in^2

Energy

1 joule (J) = 0.738 ft·lb = 2.39×10^{-4} kcal = 6.24×10^{18} eV

1 foot-pound (ft·lb) = 1.36 J = 1.29×10^{-3} Btu = 3.25×10^{-4} kcal

1 kilocalorie (kcal) = 4185 J = 3.97 Btu = 3077 ft · lb

1 Btu = 0.252 kcal = 778 ft · lb

1 kilowatthour (kW · h) = 3.6×10^6 J = 3.6 MJ

Power

1 watt (W) = 1 J/s = 0.738 ft · lb/s

1 kilowatt (kW) = 10^3 W = 1.34 hp

1 horsepower (hp) = 550 ft · lb/s = 746 W

Temperature

$$T_C = \frac{5}{9}(T_F - 32°)$$

$$T_F = \frac{9}{5} T_C + 32°$$

Appendix B

American Wire Gage Table

Resistance of bare annealed copper wire at 20°C (68°F)

AWG No.	Diameter, mils, D	Area, cmil, D^2	$\Omega/1000$ ft
0000	460.0	211 600	0.0490
000	409.6	167 800	0.0618
00	364.8	133 100	0.0779
0	324.9	105 500	0.0983
1	289.3	83 690	0.1239
2	257.6	66 360	0.1563
3	229.4	52 630	0.1970
4	204.3	41 740	0.2485
5	181.9	33 100	0.3133
6	162.0	26 250	0.3951
7	144.3	20 820	0.4982
8	128.5	16 510	0.6282
9	114.4	13 090	0.7921
10	101.9	10 380	0.9989
11	90.74	8 234	1.260
12	80.81	6 530	1.588
13	71.96	5 178	2.003
14	64.08	4 107	2.525
15	57.07	3 257	3.184
16	50.82	2 583	4.016
17	45.26	2 048	5.064
18	40.30	1 624	6.385
19	35.89	1 288	8.051
20	31.96	1 022	10.15
21	28.46	810.1	12.80
22	25.35	642.4	16.14
23	22.57	509.5	20.36
24	20.10	404.0	25.67
25	17.90	320.4	32.37
26	15.94	254.1	40.81
27	14.20	201.5	51.47
28	12.64	159.8	64.90
29	11.26	126.7	81.83
30	10.03	100.5	103.2
31	8.928	79.70	130.1
32	7.950	63.21	164.1
33	7.080	50.13	206.9
34	6.305	39.75	260.9
35	5.615	31.52	329.0
36	5.000	25.00	414.8
37	4.453	19.83	523.1
38	3.965	15.72	659.6
39	3.531	12.47	831.8
40	3.145	9.888	1049

Appendix C

Table of Allowable Current-Carrying Capacities (Ampacities) of Copper Conductors

AWG No.	Resistance, $\Omega/1000$ ft	Ampacity, A		
		Rubber-Covered, Type RH	Varnished Cambric, Type V	Asbestos, Type A
14	2.53	15	25	30
12	1.59	20	30	40
10	0.999	30	40	55
8	0.628	45	50	70
6	0.395	65	70	95
4	0.249	85	90	120
3	0.197	100	105	145
2	0.156	115	120	165
1	0.124	130	140	190
0	0.098 3	150	155	225
00	0.077 9	175	185	250
000	0.061 8	200	210	285
0000	0.049 0	230	235	340

Four-place Logarithms

N	0	1	2	3	4	5	6	7	8	9	1	2	3	4	5	6	7	8	9
													Proportional Parts						
10	0000	0043	0086	0128	0170	0212	0253	0294	0334	0374	4	8	12	17	21	25	29	33	37
11	0414	0453	0492	0531	0569	0607	0645	0682	0719	0755	4	8	11	15	19	23	26	30	34
12	0792	0828	0864	0899	0934	0969	1004	1038	1072	1106	3	7	10	14	17	21	24	28	31
13	1139	1173	1206	1239	1271	1303	1335	1367	1399	1430	3	6	10	13	16	19	23	26	29
14	1461	1492	1523	1553	1584	1614	1644	1673	1703	1732	3	6	9	12	15	18	21	24	27
15	1761	1790	1818	1847	1875	1903	1931	1959	1987	2014	3	6	8	11	14	17	20	22	25
16	2041	2068	2095	2122	2148	2175	2201	2227	2253	2279	3	5	8	11	13	16	18	21	24
17	2304	2330	2355	2380	2405	2430	2455	2480	2504	2529	2	5	7	10	12	15	17	20	22
18	2553	2577	2601	2625	2648	2672	2695	2718	2742	2765	2	5	7	9	12	14	16	19	21
19	2788	2810	2833	2856	2878	2900	2923	2945	2967	2989	2	4	7	9	11	13	16	18	20
20	3010	3032	3054	3075	3096	3118	3139	3160	3181	3201	2	4	6	8	11	13	15	17	19
21	3222	3243	3263	3284	3304	3324	3345	3365	3385	3404	2	4	6	8	10	12	14	16	18
22	3424	3444	3464	3483	3502	3522	3541	3560	3579	3598	2	4	6	8	10	12	14	15	17
23	3617	3636	3655	3674	3692	3711	3729	3747	3766	3784	2	4	6	7	9	11	13	15	17
24	3802	3820	3838	3856	3874	3892	3909	3927	3945	3962	2	4	5	7	9	11	12	14	16
25	3979	3997	4014	4031	4048	4065	4082	4099	4116	4133	2	3	5	7	9	10	12	14	15
26	4150	4166	4183	4200	4216	4232	4249	4265	4281	4298	2	3	5	7	8	10	11	13	15
27	4314	4330	4346	4362	4378	4393	4409	4425	4440	4456	2	3	5	6	8	9	11	13	14
28	4472	4487	4502	4518	4533	4548	4564	4579	4594	4609	2	3	5	6	8	9	11	12	14
29	4624	4639	4654	4669	4683	4698	4713	4728	4742	4757	1	3	4	6	7	9	10	12	13
30	4771	4786	4800	4814	4829	4843	4857	4871	4886	4900	1	3	4	6	7	9	10	11	13
31	4914	4928	4942	4955	4969	4983	4997	5011	5024	5038	1	3	4	6	7	8	10	11	12
32	5051	5065	5079	5092	5105	5119	5132	5145	5159	5172	1	3	4	5	7	8	9	11	12
33	5185	5198	5211	5224	5237	5250	5263	5276	5289	5302	1	3	4	5	6	8	9	10	12
34	5315	5328	5340	5353	5366	5378	5391	5403	5416	5428	1	3	4	5	6	8	9	10	11
35	5441	5453	5465	5478	5490	5502	5514	5527	5539	5551	1	2	4	5	6	7	9	10	11
36	5563	5575	5587	5599	5611	5623	5635	5647	5658	5670	1	2	4	5	6	7	8	10	11
37	5682	5694	5705	5717	5729	5740	5752	5763	5775	5786	1	2	3	5	6	7	8	9	10
38	5798	5809	5821	5832	5843	5855	5866	5877	5888	5899	1	2	3	5	6	7	8	9	10
39	5911	5922	5933	5944	5955	5966	5977	5988	5999	6010	1	2	3	4	5	7	8	9	10
40	6021	6031	6042	6053	6064	6075	6085	6096	6107	6117	1	2	3	4	5	6	8	9	10
41	6128	6138	6149	6160	6170	6180	6191	6201	6212	6222	1	2	3	4	5	6	7	8	9
42	6232	6243	6253	6263	6274	6284	6294	6304	6314	6325	1	2	3	4	5	6	7	8	9
43	6335	6345	6355	6365	6375	6385	6395	6405	6415	6425	1	2	3	4	5	6	7	8	9
44	6435	6444	6454	6464	6474	6484	6493	6503	6513	6522	1	2	3	4	5	6	7	8	9
45	6532	6542	6551	6561	6571	6580	6590	6599	6609	6618	1	2	3	4	5	6	7	8	9
46	6628	6637	6646	6656	6665	6675	6684	6693	6702	6712	1	2	3	4	5	6	7	7	8
47	6721	6730	6739	6749	6758	6767	6776	6785	6794	6803	1	2	3	4	5	5	6	7	8
48	6812	6821	6830	6839	6848	6857	6866	6875	6884	6893	1	2	3	4	4	5	6	7	8
49	6902	6911	6920	6928	6937	6946	6955	6964	6972	6981	1	2	3	4	4	5	6	7	8
50	6990	6998	7007	7016	7024	7033	7042	7050	7059	7067	1	2	3	3	4	5	6	7	8
51	7076	7084	7093	7101	7110	7118	7126	7135	7143	7152	1	2	3	3	4	5	6	7	8
52	7160	7168	7177	7185	7193	7202	7210	7218	7226	7235	1	2	2	3	4	5	6	7	7
53	7243	7251	7259	7267	7275	7284	7292	7300	7308	7316	1	2	2	3	4	5	6	6	7
54	7324	7332	7340	7348	7356	7364	7372	7380	7388	7396	1	2	2	3	4	5	6	6	7
N	0	1	2	3	4	5	6	7	8	9	1	2	3	4	5	6	7	8	9

Four-place Logarithms (cont.)

N	0	1	2	3	4	5	6	7	8	9	Proportional Parts 1 2 3 4 5 6 7 8 9
55	7404	7412	7419	7427	7435	7443	7451	7459	7466	7474	1 2 2 3 4 5 5 6 7
56	7482	7490	7497	7505	7513	7520	7528	7536	7543	7551	1 2 2 3 4 5 5 6 7
57	7559	7566	7574	7582	7589	7597	7604	7612	7619	7627	1 2 2 3 4 5 5 6 7
58	7634	7642	7649	7657	7664	7672	7679	7686	7694	7701	1 1 2 3 4 4 5 6 7
59	7709	7716	7723	7731	7738	7745	7752	7760	7767	7774	1 1 2 3 4 4 5 6 7
60	7782	7789	7796	7803	7810	7818	7825	7832	7839	7846	1 1 2 3 4 4 5 6 6
61	7853	7860	7868	7875	7882	7889	7896	7903	7910	7917	1 1 2 3 4 4 5 6 6
62	7924	7931	7938	7945	7952	7959	7966	7973	7980	7987	1 1 2 3 3 4 5 6 6
63	7993	8000	8007	8014	8021	8028	8035	8041	8048	8055	1 1 2 3 3 4 5 5 6
64	8062	8069	8075	8082	8089	8096	8102	8109	8116	8122	1 1 2 3 3 4 5 5 6
65	8129	8136	8142	8149	8156	8162	8169	8176	8182	8189	1 1 2 3 3 4 5 5 6
66	8195	8202	8209	8215	8222	8228	8235	8241	8248	8254	1 1 2 3 3 4 5 5 6
67	8261	8267	8274	8280	8287	8293	8299	8306	8312	8319	1 1 2 3 3 4 5 5 6
68	8325	8331	8338	8344	8351	8357	8363	8370	8376	8382	1 1 2 3 3 4 4 5 6
69	8388	8395	8401	8407	8414	8420	8426	8432	8439	8445	1 1 2 2 3 4 4 5 6
70	8451	8457	8463	8470	8476	8482	8488	8494	8500	8506	1 1 2 2 3 4 4 5 6
71	8513	8519	8525	8531	8537	8543	8549	8555	8561	8567	1 1 2 2 3 4 4 5 5
72	8573	8579	8585	8591	8597	8603	8609	8615	8621	8627	1 1 2 2 3 4 4 5 5
73	8633	8639	8645	8651	8657	8663	8669	8675	8681	8686	1 1 2 2 3 4 4 5 5
74	8692	8698	8704	8710	8716	8722	8727	8733	8739	8745	1 1 2 2 3 4 4 5 5
75	8751	8756	8762	8768	8774	8779	8785	8791	8797	8802	1 1 2 2 3 3 4 5 5
76	8808	8814	8820	8825	8831	8837	8842	8848	8854	8859	1 1 2 2 3 3 4 5 5
77	8865	8871	8876	8882	8887	8893	8899	8904	8910	8915	1 1 2 2 3 3 4 4 5
78	8921	8927	8932	8938	8943	8949	8954	8960	8965	8971	1 1 2 2 3 3 4 4 5
79	8976	8982	8987	8993	8998	9004	9009	9015	9020	9025	1 1 2 2 3 3 4 4 5
80	9031	9036	9042	9047	9053	9058	9063	9069	9074	9079	1 1 2 2 3 3 4 4 5
81	9085	9090	9096	9101	9106	9112	9117	9122	9128	9133	1 1 2 2 3 3 4 4 5
82	9138	9143	9149	9154	9159	9165	9170	9175	9180	9186	1 1 2 2 3 3 4 4 5
83	9191	9196	9201	9206	9212	9217	9222	9227	9232	9238	1 1 2 2 3 3 4 4 5
84	9243	9248	9253	9258	9263	9269	9274	9279	9284	9289	1 1 2 2 3 3 4 4 5
85	9294	9299	9304	9309	9315	9320	9325	9330	9335	9340	1 1 2 2 3 3 4 4 5
86	9345	9350	9355	9360	9365	9370	9375	9380	9385	9390	1 1 2 2 3 3 4 4 5
87	9395	9400	9405	9410	9415	9420	9425	9430	9435	9440	0 1 1 2 2 3 3 4 4
88	9445	9450	9455	9460	9465	9469	9474	9479	9484	9489	0 1 1 2 2 3 3 4 4
89	9494	9499	9504	9509	9513	9518	9523	9528	9533	9538	0 1 1 2 2 3 3 4 4
90	9542	9547	9552	9557	9562	9566	9571	9576	9581	9586	0 1 1 2 2 3 3 4 4
91	9590	9595	9600	9605	9609	9614	9619	9624	9628	9633	0 1 1 2 2 3 3 4 4
92	9638	9643	9647	9652	9657	9661	9666	9671	9675	9680	0 1 1 2 2 3 3 4 4
93	9685	9689	9694	9699	9703	9708	9713	9717	9722	9727	0 1 1 2 2 3 3 4 4
94	9731	9736	9741	9745	9750	9754	9759	9763	9768	9773	0 1 1 2 2 3 3 4 4
95	9777	9782	9786	9791	9795	9800	9805	9809	9814	9818	0 1 1 2 2 3 3 4 4
96	9823	9827	9832	9836	9841	9845	9850	9854	9859	9863	0 1 1 2 2 3 3 4 4
97	9868	9872	9877	9881	9886	9890	9894	9899	9903	9908	0 1 1 2 2 3 3 4 4
98	9912	9917	9921	9926	9930	9934	9939	9943	9948	9952	0 1 1 2 2 3 3 4 4
99	9956	9961	9965	9969	9974	9978	9983	9987	9991	9996	0 1 1 2 2 3 3 3 4
N	0	1	2	3	4	5	6	7	8	9	1 2 3 4 5 6 7 8 9

Antilogarithms

	0	1	2	3	4	5	6	7	8	9	1	2	3	4	5	6	7	8	9
.00	1000	1002	1005	1007	1009	1012	1014	1016	1019	1021	0	0	1	1	1	1	2	2	2
.01	1023	1026	1028	1030	1033	1035	1038	1040	1042	1045	0	0	1	1	1	1	2	2	2
.02	1047	1050	1052	1054	1057	1059	1062	1064	1067	1069	0	0	1	1	1	1	2	2	2
.03	1072	1074	1076	1079	1081	1084	1086	1089	1091	1094	0	0	1	1	1	1	2	2	2
.04	1096	1099	1102	1104	1107	1109	1112	1114	1117	1119	0	1	1	1	1	2	2	2	2
.05	1122	1125	1127	1130	1132	1135	1138	1140	1143	1146	0	1	1	1	1	2	2	2	2
.06	1148	1151	1153	1156	1159	1161	1164	1167	1169	1172	0	1	1	1	1	2	2	2	2
.07	1175	1178	1180	1183	1186	1189	1191	1194	1197	1199	0	1	1	1	1	2	2	2	2
.08	1202	1205	1208	1211	1213	1216	1219	1222	1225	1227	0	1	1	1	1	2	2	2	3
.09	1230	1233	1236	1239	1242	1245	1247	1250	1253	1256	0	1	1	1	1	2	2	2	3
.10	1259	1262	1265	1268	1271	1274	1276	1279	1282	1285	0	1	1	1	1	2	2	2	3
.11	1288	1291	1294	1297	1300	1303	1306	1309	1312	1315	0	1	1	1	2	2	2	2	3
.12	1318	1321	1324	1327	1330	1334	1337	1340	1343	1346	0	1	1	1	2	2	2	3	3
.13	1349	1352	1355	1358	1361	1365	1368	1371	1374	1377	0	1	1	1	2	2	2	3	3
.14	1380	1384	1387	1390	1393	1396	1400	1403	1406	1409	0	1	1	1	2	2	2	3	3
.15	1413	1416	1419	1422	1426	1429	1432	1435	1439	1442	0	1	1	1	2	2	2	3	3
.16	1445	1449	1452	1455	1459	1462	1466	1469	1472	1476	0	1	1	1	2	2	2	3	3
.17	1479	1483	1486	1489	1493	1496	1500	1503	1507	1510	0	1	1	1	2	2	2	3	3
.18	1514	1517	1521	1524	1528	1531	1535	1538	1542	1545	0	1	1	1	2	2	2	3	3
.19	1549	1552	1556	1560	1563	1567	1570	1574	1578	1581	0	1	1	1	2	2	3	3	3
.20	1585	1589	1592	1596	1600	1603	1607	1611	1614	1618	0	1	1	1	2	2	3	3	3
.21	1622	1626	1629	1633	1637	1641	1644	1648	1652	1656	0	1	1	2	2	2	3	3	3
.22	1660	1663	1667	1671	1675	1679	1683	1687	1690	1694	0	1	1	2	2	2	3	3	3
.23	1698	1702	1706	1710	1714	1718	1722	1726	1730	1734	0	1	1	2	2	2	3	3	4
.24	1738	1742	1746	1750	1754	1758	1762	1766	1770	1774	0	1	1	2	2	2	3	3	4
.25	1778	1782	1786	1791	1795	1799	1803	1807	1811	1816	0	1	1	2	2	2	3	3	4
.26	1820	1824	1828	1832	1837	1841	1845	1849	1854	1858	0	1	1	2	2	3	3	3	4
.27	1862	1866	1871	1875	1879	1884	1888	1892	1897	1901	0	1	1	2	2	3	3	3	4
.28	1905	1910	1914	1919	1923	1928	1932	1936	1941	1945	0	1	1	2	2	3	3	4	4
.29	1950	1954	1959	1963	1968	1972	1977	1982	1986	1991	0	1	1	2	2	3	3	4	4
.30	1995	2000	2004	2009	2014	2018	2023	2028	2032	2037	0	1	1	2	2	3	3	4	4
.31	2042	2046	2051	2056	2061	2065	2070	2075	2080	2084	0	1	1	2	2	3	3	4	4
.32	2089	2094	2099	2104	2109	2113	2118	2123	2128	2133	0	1	1	2	2	3	3	4	4
.33	2138	2143	2148	2153	2158	2163	2168	2173	2178	2183	0	1	1	2	2	3	3	4	4
.34	2188	2193	2198	2203	2208	2213	2218	2223	2228	2234	1	1	2	2	3	3	4	4	5
.35	2239	2244	2249	2254	2259	2265	2270	2275	2280	2286	1	1	2	2	3	3	4	4	5
.36	2291	2296	2301	2307	2312	2317	2323	2328	2333	2339	1	1	2	2	3	3	4	4	5
.37	2344	2350	2355	2360	2366	2371	2377	2382	2388	2393	1	1	2	2	3	3	4	4	5
.38	2399	2404	2410	2415	2421	2427	2432	2438	2443	2449	1	1	2	2	3	3	4	4	5
.39	2455	2460	2466	2472	2477	2483	2489	2495	2500	2506	1	1	2	2	3	3	4	5	5
.40	2512	2518	2523	2529	2535	2541	2547	2553	2559	2564	1	1	2	2	3	4	4	5	5
.41	2570	2576	2582	2588	2594	2600	2606	2612	2618	2624	1	1	2	2	3	4	4	5	5
.42	2630	2636	2642	2649	2655	2661	2667	2673	2679	2685	1	1	2	2	3	4	4	5	6
.43	2692	2698	2704	2710	2716	2723	2729	2735	2742	2748	1	1	2	3	3	4	4	5	6
.44	2754	2761	2767	2773	2780	2786	2793	2799	2805	2812	1	1	2	3	3	4	4	5	6
.45	2818	2825	2831	2838	2844	2851	2858	2864	2871	2877	1	1	2	3	3	4	5	5	6
.46	2884	2891	2897	2904	2911	2917	2924	2931	2938	2944	1	1	2	3	3	4	5	5	6
.47	2951	2958	2965	2972	2979	2985	2992	2999	3006	3013	1	1	2	3	3	4	5	5	6
.48	3020	3027	3034	3041	3048	3055	3062	3069	3076	3083	1	1	2	3	4	4	5	6	6
.49	3090	3097	3105	3112	3119	3126	3133	3141	3148	3155	1	1	2	3	4	4	5	6	6
	0	1	2	3	4	5	6	7	8	9	1	2	3	4	5	6	7	8	9

Antilogarithms (cont.)

	0	1	2	3	4	5	6	7	8	9	1	2	3	4	5	6	7	8	9
.50	3162	3170	3177	3184	3192	3199	3206	3214	3221	3228	1	1	2	3	4	4	5	6	7
.51	3236	3243	3251	3258	3266	3273	3281	3289	3296	3304	1	2	2	3	4	5	5	6	7
.52	3311	3319	3327	3334	3342	3350	3357	3365	3373	3381	1	2	2	3	4	5	5	6	7
.53	3388	3396	3404	3412	3420	3428	3436	3443	3451	3459	1	2	2	3	4	5	6	6	7
.54	3467	3475	3483	3491	3499	3508	3516	3524	3532	3540	1	2	2	3	4	5	6	6	7
.55	3548	3556	3565	3573	3581	3589	3597	3606	3614	3622	1	2	2	3	4	5	6	7	7
.56	3631	3639	3648	3656	3664	3673	3681	3690	3698	3707	1	2	3	3	4	5	6	7	8
.57	3715	3724	3733	3741	3750	3758	3767	3776	3784	3793	1	2	3	3	4	5	6	7	8
.58	3802	3811	3819	3828	3837	3846	3855	3864	3873	3882	1	2	3	4	4	5	6	7	8
.59	3890	3899	3908	3917	3926	3936	3945	3954	3963	3972	1	2	3	4	5	5	6	7	8
.60	3981	3990	3999	4009	4018	4027	4036	4046	4055	4064	1	2	3	4	5	6	6	7	8
.61	4074	4083	4093	4102	4111	4121	4130	4140	4150	4159	1	2	3	4	5	6	7	8	9
.62	4169	4178	4188	4198	4207	4217	4227	4236	4246	4256	1	2	3	4	5	6	7	8	9
.63	4266	4276	4285	4295	4305	4315	4325	4335	4345	4355	1	2	3	4	5	6	7	8	9
.64	4365	4375	4385	4395	4406	4416	4426	4436	4446	4457	1	2	3	4	5	6	7	8	9
.65	4467	4477	4487	4498	4508	4519	4529	4539	4550	4560	1	2	3	4	5	6	7	8	9
.66	4571	4581	4592	4603	4613	4624	4634	4645	4656	4667	1	2	3	4	5	6	7	9	10
.67	4677	4688	4699	4710	4721	4732	4742	4753	4764	4775	1	2	3	4	5	7	8	9	10
.68	4786	4797	4808	4819	4831	4842	4853	4864	4875	4887	1	2	3	4	6	7	8	9	10
.69	4898	4909	4920	4932	4943	4955	4966	4977	4989	5000	1	2	3	5	6	7	8	9	10
.70	5012	5023	5035	5047	5058	5070	5082	5093	5105	5117	1	2	4	5	6	7	8	9	11
.71	5129	5140	5152	5164	5176	5188	5200	5212	5224	5236	1	2	4	5	6	7	8	10	11
.72	5248	5260	5272	5284	5297	5309	5321	5333	5346	5358	1	2	4	5	6	7	9	10	11
.73	5370	5383	5395	5408	5420	5433	5445	5458	5470	5483	1	3	4	5	6	8	9	10	11
.74	5495	5508	5521	5534	5546	5559	5572	5585	5598	5610	1	3	4	5	6	8	9	10	12
.75	5623	5636	5649	5662	5675	5689	5702	5715	5728	5741	1	3	4	5	7	8	9	10	12
.76	5754	5768	5781	5794	5808	5821	5834	5848	5861	5875	1	3	4	5	7	8	9	11	12
.77	5888	5902	5916	5929	5943	5957	5970	5984	5998	6012	1	3	4	5	7	8	10	11	12
.78	6026	6039	6053	6067	6081	6095	6109	6124	6138	6152	1	3	4	6	7	8	10	11	13
.79	6166	6180	6194	6209	6223	6237	6252	6266	6281	6295	1	3	4	6	7	9	10	11	13
.80	6310	6324	6339	6353	6368	6383	6397	6412	6427	6442	1	3	4	6	7	9	10	12	13
.81	6457	6471	6486	6501	6516	6531	6546	6561	6577	6592	2	3	5	6	8	9	11	12	14
.82	6607	6622	6637	6653	6668	6683	6699	6714	6730	6745	2	3	5	6	8	9	11	12	14
.83	6761	6776	6792	6808	6823	6839	6855	6871	6887	6902	2	3	5	6	8	9	11	13	14
.84	6918	6934	6950	6966	6982	6998	7015	7031	7047	7063	2	3	5	6	8	10	11	13	15
.85	7079	7096	7112	7129	7145	7161	7178	7194	7211	7228	2	3	5	7	8	10	12	13	15
.86	7244	7261	7278	7295	7311	7328	7345	7362	7379	7396	2	3	5	7	8	10	12	13	15
.87	7413	7430	7447	7464	7482	7499	7516	7534	7551	7568	2	3	5	7	9	10	12	14	16
.88	7586	7603	7621	7638	7656	7674	7691	7709	7727	7745	2	4	5	7	9	11	12	14	16
.89	7762	7780	7798	7816	7834	7852	7870	7889	7907	7925	2	4	5	7	9	11	13	14	16
.90	7943	7962	7980	7998	8017	8035	8054	8072	8091	8110	2	4	6	7	9	11	13	15	17
.91	8128	8147	8166	8185	8204	8222	8241	8260	8279	8299	2	4	6	8	9	11	13	15	17
.92	8318	8337	8356	8375	8395	8414	8433	8453	8472	8492	2	4	6	8	10	12	14	15	17
.93	8511	8531	8551	8570	8590	8610	8630	8650	8670	8690	2	4	6	8	10	12	14	16	18
.94	8710	8730	8750	8770	8790	8810	8831	8851	8872	8892	2	4	6	8	10	12	14	16	18
.95	8913	8933	8954	8974	8995	9016	9036	9057	9078	9099	2	4	6	8	10	12	15	17	19
.96	9120	9141	9162	9183	9204	9226	9247	9268	9290	9311	2	4	6	8	11	13	15	17	19
.97	9333	9354	9376	9397	9419	9441	9462	9484	9506	9528	2	4	7	9	11	13	15	17	20
.98	9550	9572	9594	9616	9638	9661	9683	9705	9727	9750	2	4	7	9	11	13	16	18	20
.99	9772	9795	9817	9840	9863	9886	9908	9931	9954	9977	2	5	7	9	11	14	16	18	20
	0	1	2	3	4	5	6	7	8	9	1	2	3	4	5	6	7	8	9

Appendix E

Natural Trigonometric Functions

Angle Deg.	Angle Rad.	Sin	Cos	Tan	Angle Deg.	Angle Rad.	Sin	Cos	Tan
0°	0.000	0.000	1.000	0.000					
1°	0.017	0.018	1.000	0.018	46°	0.803	0.719	0.695	1.036
2°	0.035	0.035	0.999	0.035	47°	0.820	0.731	0.682	1.072
3°	0.052	0.052	0.999	0.052	48°	0.838	0.743	0.669	1.111
4°	0.070	0.070	0.998	0.070	49°	0.855	0.755	0.656	1.150
5°	0.087	0.087	0.996	0.088	50°	0.873	0.766	0.643	1.192
6°	0.105	0.105	0.995	0.105	51°	0.890	0.777	0.629	1.235
7°	0.122	0.122	0.993	0.123	52°	0.908	0.788	0.616	1.280
8°	0.140	0.139	0.990	0.141	53°	0.925	0.799	0.602	1.327
9°	0.157	0.156	0.988	0.158	54°	0.942	0.809	0.588	1.376
10°	0.175	0.174	0.985	0.176	55°	0.960	0.819	0.574	1.428
11°	0.192	0.191	0.982	0.194	56°	0.977	0.829	0.559	1.483
12°	0.209	0.208	0.978	0.213	57°	0.995	0.839	0.545	1.540
13°	0.227	0.225	0.974	0.231	58°	1.012	0.848	0.530	1.600
14°	0.244	0.242	0.970	0.249	59°	1.030	0.857	0.515	1.664
15°	0.262	0.259	0.966	0.268	60°	1.047	0.866	0.500	1.732
16°	0.279	0.276	0.961	0.287	61°	1.065	0.875	0.485	1.804
17°	0.297	0.292	0.956	0.306	62°	1.082	0.883	0.470	1.881
18°	0.314	0.309	0.951	0.325	63°	1.100	0.891	0.454	1.963
19°	0.332	0.326	0.946	0.344	64°	1.117	0.899	0.438	2.050
20°	0.349	0.342	0.940	0.364	65°	1.134	0.906	0.423	2.145
21°	0.367	0.358	0.934	0.384	66°	1.152	0.914	0.407	2.246
22°	0.384	0.375	0.927	0.404	67°	1.169	0.921	0.391	2.356
23°	0.401	0.391	0.921	0.425	68°	1.187	0.927	0.375	2.475
24°	0.419	0.407	0.914	0.445	69°	1.204	0.934	0.358	2.605
25°	0.436	0.423	0.906	0.466	70°	1.222	0.940	0.342	2.747
26°	0.454	0.438	0.899	0.488	71°	1.239	0.946	0.326	2.904
27°	0.471	0.454	0.891	0.510	72°	1.257	0.951	0.309	3.078
28°	0.489	0.470	0.883	0.532	73°	1.274	0.956	0.292	3.271
29°	0.506	0.485	0.875	0.554	74°	1.292	0.961	0.276	3.487
30°	0.524	0.500	0.866	0.577	75°	1.309	0.966	0.259	3.732
31°	0.541	0.515	0.857	0.601	76°	1.326	0.970	0.242	4.011
32°	0.559	0.530	0.848	0.625	77°	1.344	0.974	0.225	4.331
33°	0.576	0.545	0.839	0.649	78°	1.361	0.978	0.208	4.705
34°	0.593	0.559	0.829	0.675	79°	1.379	0.982	0.191	5.145
35°	0.611	0.574	0.819	0.700	80°	1.396	0.985	0.174	5.671
36°	0.628	0.588	0.809	0.727	81°	1.414	0.988	0.156	6.314
37°	0.646	0.602	0.799	0.754	82°	1.431	0.990	0.139	7.115
38°	0.663	0.616	0.788	0.781	83°	1.449	0.993	0.122	8.144
39°	0.681	0.629	0.777	0.810	84°	1.466	0.995	0.105	9.514
40°	0.698	0.643	0.766	0.839	85°	1.484	0.996	0.087	11.43
41°	0.716	0.658	0.755	0.869	86°	1.501	0.998	0.070	14.30
42°	0.733	0.669	0.743	0.900	87°	1.518	0.999	0.052	19.08
43°	0.751	0.682	0.731	0.933	88°	1.536	0.999	0.035	28.64
44°	0.768	0.695	0.719	0.966	89°	1.553	1.000	0.018	57.29
45°	0.785	0.707	0.707	1.000	90°	1.571	1.000	0.000	∞

INDEX